# Concise Handbook of Waste Treatment Technologies

# Concise Handbook of Waste Treatment Technologies

Saleh S. Al Arni and
Mahmoud M. Elwaheidi

**CRC Press**
Taylor & Francis Group
Boca Raton London New York

CRC Press is an imprint of the
Taylor & Francis Group, an informa business

First edition published 2021
by CRC Press
6000 Broken Sound Parkway NW, Suite 300, Boca Raton, FL 33487-2742

and by CRC Press
2 Park Square, Milton Park, Abingdon, Oxon OX14 4RN

© 2021 Taylor & Francis Group, LLC

CRC Press is an imprint of Taylor & Francis Group, LLC

*Library of Congress Cataloging-in-Publication Data*
A catalog record has been requested for this book

ISBN: 978-0-367-63130-7 (hbk)
ISBN: 978-0-367-63129-1 (pbk)
ISBN: 978-1-003-13664-4 (ebk)

*We dedicate this work to the memory of our parents with great affection.*

# Contents

# *Preface*

Production of waste material is an inevitable outcome of our daily life activities. These materials could have harmful effects on our health, environment and planet. Therefore, efficient processing and management technologies are deemed necessary to get rid of these materials in a wise way. There are different kinds of waste treatment technologies that are associated with different kinds of waste products.

An overall review of the available textbooks on waste treatment reveals the use of a non-integrated approach that deals separately with each type of waste materials (i.e. solids, liquids and gases). This work comes to satisfy a part of the actual need for a concise and comprehensive guidebook that deals with the contemporary waste treatment technologies in a global and integrated approach.

The "Concise Handbook of Waste Treatment Technologies" is intended to provide one single handbook that deals with various treatment processes that can be applied to all waste types. It, concurrently, provides an adequate didactic textbook for undergraduate students in various engineering disciplines (civil, chemical, mechanical, sanitary, public health and environmental) and professionals who are interested in modern waste treatment technologies and management. In addition, this book will provide the ordinary reader with a better understanding of how waste materials are treated and managed in a manner compatible with the engineering, health, safety and environmental regulations and laws.

This book comes in 12 chapters that cover the intended objectives. Chapter 1 provides general information about waste, waste management and disposal. The impact of waste on the health, safety and environment is dealt with in Chapter 2. Waste related laws, regulations and standards are discussed in Chapter 3. The classification of waste materials is covered in Chapter 4. Waste management hierarchy and solid waste management are discussed in Chapters 5 and 6, respectively. In Chapter 7, waste treatment processes are illustrated and explained. In Chapters 8 and 9, biowaste solid materials treatment and waste disposal by thermal processes are discussed, respectively. Industrial wastewater treatment and effluent gas control (clean-up system) are dealt with in Chapters 10 and 11, respectively. The economics of waste treatment and management are emphasized in Chapter 12.

For suggestions, comments or inquiries regarding this book, kindly email the authors at:

*Dr. Saleh S. Al Arni, PhD*
Chemical Engineering Department, Hail University, KSA
arnisaleh@hotmail.com
*Dr. Mahmoud M. Elwaheidi, PhD*
Geology and Geophysics Department, King Saud University, KSA
melwaheidi@ksu.edu.sa

# Author Biographies

 Saleh S. Al Arni is a Professor at Chemical Engineering Department, Hail University, (UOH), Kingdom of Saudi Arabia. He received the master degree (Laurea) in Chemical Engineering from Faculty of Engineering, University of Genoa (Italy), in January 1996. He is specialized in field of "Technological Application of Fermentation Processes". In February 2000, he received the first PhD degree in "Technologies and Economic of the Processes and Products to Safeguard the Environment" from Catania University (Italy), and in April 2008, he received the second PhD degree in "Chemical Sciences, Technologies and Processes", from Genoa University (Italy).

Since his graduation, he has a strong enthusiasm and interest in research and teaching activities. He graduated in 1996 and joined, in the same year, the Biotechnology and Agroindustrial Technologies Group in the "Department of Chemical and Process Engineering" at Genoa University (Italy). His research and teaching activities deal with biotechnological processes. During his academic career, he worked at several positions.

He published several scientific articles in international journals (Scholar; Research gate, Thomson Reuters, ORCID and Scopus) and two scientific book chapters.

He supervised several theses, and act as a peer reviewer for many International Journals such as Bioresource Technology, Fuel Processing Technology, Hydrogen Energy, Journal of King Saud University, Engineering Sciences, Journal of Geography and Regional Planning, Environmental Engineering and Management Journal, International Journal of Biotechnology and Biomaterials Engineering, Revista Maxicana de Ingenieria Quimica, Energy Conversion and Management (ECM), Reviews in Chemical Engineering, Industrial Crops and Products, Iranian Journal of Chemistry and Chemical Engineering (IJCCE) and Nanoscience & Nanotechnology-Asia.

Dr. Mahmoud M. Elwaheidi is a Professor at the Department of Geology & Geophysics, King Saud University, Riyadh (KSA). He was awarded his PhD in Environmental & Engineering Geophysics from the University of Genova, Genova (Italy) in 1994. He received his M.Sc. in Applied Geophysics from the University of Jordan, Amman (Jordan). In 1987, he received his B.Sc. in Earth and Environmental Sciences from the University of Yarmouk, Irbid (Jordan).

Dr. Elwaheidi has more than 20 years of professional and academic experience in the fields of applied geosciences. He is interested in research topics that address environmental and engineering issues including Health Safety and Environmental Impact Assessment (HSEIA) and climate change studies.

He published several scientific publications in ISI journals and acting as a scientific reviewer and board member of several specialized prestigious journals. Dr. Elwaheidi is also a scientific writer and a translator certified by several UN agencies that deal with natural resources and environmental issues.

# 1

## General Concept of Waste Management

**Key Learning Objectives**

- Understanding the terminology used in waste management.
- Understanding the basic concept of technologies used for waste treatment.

## 1.1 Introduction

This chapter briefly discusses the concepts of waste, waste treatment technologies, nature of waste and the ways it is produced. There is a strong relationship between waste generation and urbanization. The generated waste must be treated to reduce its hazardous health, safety and environmental effects. Waste treatment depends on a set of organizational, structural and technical measures. It also depends on the available economic means and the adapted waste management policies. The scope of waste management is to delay natural resource consumption by applying certain practices starting from resource points and extending to recovery stations. The main necessity is to enrich the resources that are depleted due to rising population and higher consumption rates.

## 1.2 Terminology

Variety, quality and quantity of waste materials imposed the use of various terminologies that will be dealt with in detail in the chapters of this book.

The term *waste* is commonly used in any human activity. The use of this term becomes more complicated due to the enormous technological and industrial developments that continually add new products, producing new types of waste. However, the term waste can be defined as unusable or discarded materials sorted from human activity and intended to be disposed of anyway.

Literature contains many terms associated with waste such as landfill, compost, refuse, garbage, dust and litter. Oxford Advanced Learner's Dictionary makes a clear distinction between *rubbish*, *garbage*, *trash* and *refuse*:

> *Rubbish* is the usual word in British English for the things that you throw away because you no longer want or need them. *Garbage* and *trash* are both used in North American English. Inside the home, *garbage* tends to mean waste food and other wet material, while *trash* is paper, cardboard and dry material. In British English, you put your rubbish in a dustbin in the street to be collected by the dustmen. In North American English, your garbage and trash go in a garbage/trashcan in the street and is collected by garbage men/collectors. *Refuse* is a formal word and is used in both British English and North American English. Refuse collector is the formal word for a dustman or garbage collector.[1]

Definition of waste in Basel Convention* (Article 2, paragraph 1) is "substances or objects which are disposed of or are intended to be disposed of or are required to be disposed of by the provisions of national law" [2].

Also, *Junk* word refers to things that are considered useless or of little value, and *litter* word refers to small pieces of rubbish or garbage such as paper, cans and bottles that people have left lying in the public place; in other words, these are the wastes that are not put in the correct bin.

The term *waste management* is related to the process of monitoring waste materials starting from minimization/prevention through collection, transport, recycling, treatment and disposal. This includes all types of waste materials such as solids, liquids, gaseous and radioactive waste.

Waste treatment includes several activities such as organizational, structural and technical waste treatment measures, controlled landfilling, thermal treatment, biological treatment, recycling mechanical treatment and economic aspect. Local government is responsible for managing nonhazardous residential, commercial and institutional waste materials. However, managing hazardous and industrial waste materials is the generator's responsibility, subject to local laws.

The term *disposal* refers to intentional burial, deposit, discharge, dumping, placing or release of any waste material into or on any air, land or water.

---

* The Basel Convention is an international regulation for transboundary movements of hazardous materials and other wastes between nations. It was signed on 22 March 1989, and has entered into force on 5 May 1992, in July 2011 there were 175 parties.

Disposal in Basel Convention (Annex IVB of the Basel Convention) means "any operation which may lead to resource recovery, recycling, reclamation, direct re-use or alternative uses".

In this book, the term waste refers to an unusable or unwanted material that is sorted from human activity and intended to be disposed of anyway. In literature, different words such as "toxic, poisonous, chemical, and special" were used to refer to waste. We use the term *hazardous waste*, which means any waste or materials that pose a threat to human health and/or the environment; typically, this covers all types of hazardous materials including radioactive waste. The hazardous waste must be treated and disposed of separately from nonhazardous waste.

## 1.3 This Book

This book discusses waste and waste treatment methods. In general, the scope of waste management is to reduce the effect of waste on human health, safety, environment or aesthetics. This will be discussed in Chapter 2. Reducing waste material effects depends on their severity and should be dealt with by national and international laws, and this topic will be discussed in Chapter 3. Waste treatment processes depend on the waste categories and the technologies applied. There are many categories of waste, for example, solid, liquid, gaseous, hazardous, nonhazardous and degradable; these will be discussed in Chapter 4.

Chapter 5 will discuss waste management hierarchy. Waste processing and treatment includes many different processes, including mechanical, thermal and biological processes. Each technology depends on the sub-treatment of waste materials. Chapter 6 will address these processes and waste disposal methods.

Some specific industrial waste needs special treatment; any treatment depends on several parameters, including state of the material. Industrial solid, industrial liquid and industrial effluent gas will be discussed in Chapters 7–11. Chapter 12 will discuss the economic impact of waste treatment.

## Review Questions

Define the following terms:
Waste
waste treatment process
waste disposal

## References

1. Oxford Advanced Learner's Dictionary, new edition online, http://oald8. oxfordlearnersdictionaries.com.
2. Basel Convention: Instruction manual on the prosecution of illegal traffic of hazardous waste or other waste, 2012, the United Nations Environment Program (UNEP), www.basel.int.

# 2

## Waste Impact on Health and Environment

### Key Learning Objective

- Understanding the impacts of waste on human health, safety and the environment.

## 2.1 Introduction

Waste materials cause pollutions in air, land and water. There are many different substances emitted from waste that consists principally of methane and carbon dioxide, hydrogen sulfides, a mixture of volatile organic compounds (VOCs) such as polycyclic aromatic hydrocarbons, metals such as mercury vapor, pesticides and pathogens. These pollutants should be classified based on their degree of toxicity or other hazards and their persistence in the environment. Thus, waste management (such as the collection, transport processing and disposal of waste) is considered a very critical issue that affects public health.

## 2.2 Impacts of Waste on Human Health

Waste materials that are generated by hospitals, health care centers, medical laboratories and research centers require special treatment since they could be sources of major health hazards to human health, for example, infections and spreading of diseases.

Health and environmental impacts of waste are associated with the release of traces of gases arising from the waste materials that were often placed in accumulation. In fact, these materials decompose and release a bad odor and create unhygienic conditions. Improper management and disposal of waste materials cause serious health problems to the environment. In addition, the improper accumulation of waste attracts flies, rats and other species that spread infectious diseases.

The risks of infection and injury are increased due to the uncollected waste materials, which also result in damages to the environment. The fermentation of organic waste materials produces leachate, which provides conditions that are favorable to the growth of microbial pathogens; hence, various types of infections and diseases may occur.

The population that are most vulnerable to hazardous waste materials are preschool children and waste workers. In addition, neighborhoods of waste dump areas are more vulnerable to aforementioned pollutants. Direct exposure to chemicals can lead to diseases by chemical poisoning. Many studies have been carried out on the connection between health and hazardous waste [1]. The impact caused by a pollutant material depends on the individual's health status and the degree of exposure [2]. Figure 2.1 illustrates the various health impacts of pollution on the human being.

Waste materials have a dangerous impact on soil and water sources because the contaminant materials block water runoff that causes the formation of

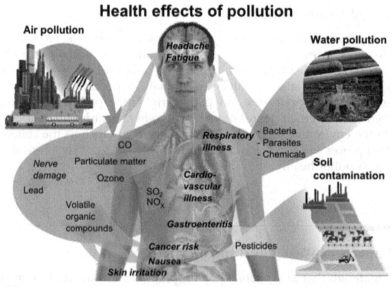

**FIGURE 2.1**

An illustration showing the impacts of pollution on human health (Pollution: What Autism, ADD, and Dyslexia Have in Common; Posted on December 1, 2011 by Tafline Laylin in Health with 4 Comments. See more at: www.greenprophet.com/2011/12/pollution-autism-add-dyslexia/#sthash.IotnUk1p.dpuf, available on 12/6/2020).

**FIGURE 2.2**

Toxic liquid, leachate, is coming out of a pipe immersed in solid waste (S. Last, The Landfill Leachate News Blog, http://leachate.blogspot.com/p/what-is-leachate-landfill-leachaet.html, available on 12/6/2020).

(a)           (b)

**FIGURE 2.3**

Pollution impact of waste on water sources: (a) rivers (Sometimes Interesting, http:// sometimes-interesting.com/2011/07/17/electronic-waste-dump-of-the-world/, available on line at 26/6/2014) and (b) seas (Waste and the Sustainable Development Goals by Z. Lenkiewicz, May 11, 2016; https://wasteaid.org/waste-sustainable-development-goals/, available on line at 12/6/2020).

stagnant water pools that become disease breeding grounds. Furthermore, contamination of water bodies and ground water sources is caused by direct dumping of waste materials (Figure 2.2). Decomposed solid waste materials generate toxic liquid (leachate) that has hazardous impacts on human health. Figure 2.3 shows examples on pollution impacts of waste on water sources.

Industrial hazardous waste materials and municipal waste materials should be disposed separately to avoid radioactive and chemical hazards. Waste disposal site impacts include litter, odors, flies, animals and birds. Figure 2.4 displays some of those impacts.

|        |        |        |
|--------|--------|--------|
| (a)    | (b)    | (c)    |

**FIGURE 2.4**

Impact of waste disposal sites on (a) human (Environment Insider, http://environmentinsider.com/poor-waste-management-effects/, available on line at 26/6/2014), (b) animals (homesinthecity.org; http://homesinthecity.org/en/hic-fellowships/solid-waste-management/, available on line at 12/6/2020) and (c) birds (R. Altaf, Jan 14, 2019, Environment, Garbage Dumps Are Changing the Food Habits of Wild Animals; https://thewire.in/environment/garbage-dumps-are-changing-the-food-habits-of-wild-animals, available on line at 12/6/2020).

## 2.3 Occupational Hazards Associated with Waste Handling

Occupational health issues of the workforce involved in waste management is serious and must be taken into consideration. In fact, the tasks to be carried out by a waste treatment and management worker vary between waste collection, sorting and recycling in different locations such as streets, collection stations, landfill sites and incineration stations. Truck drivers and laborers employed in waste collection, sorting and disposal may be exposed to the potential hazards of wastes. This is mainly because they are exposed to infections and chronic diseases such as bio aerosols and volatile compounds related to respiratory, gastrointestinal, skin problems and cancer diseases. Also, waste collection workers suffer from high rates of occupational accidents such as poisoning and chemical burns. Some examples are provided below [3].

### 2.3.1 Infections

- Contact with waste materials cause infections of skin and blood.
- Infections of eyes and respiratory system caused by exposure to dust during working in a landfill.
- Flies feeding on the waste materials cause infections of intestines.
- Animals feeding on waste materials transmit diseases.

### 2.3.2 Chronic Diseases

- Exposure to hazardous compounds and dust could lead to suffering from chronic respiratory diseases.

### 2.3.3 Accidents

- Dealing with heavy weights could cause bone and muscle problems.
- Contact with sharp objects could cause wound infections.
- Dealing with hazardous chemical waste materials could result in poisoning and chemical burns.
- Accidents occur at waste disposal sites that cause several types of injuries.

## 2.4 Impacts of Waste on the Environment (Air Pollution)

Waste treatment and disposal sites can create health hazards, for example, improper operations of waste incineration of clinical, chemical, sewage sludge and municipal waste produces a large number of air pollutants (Figure 2.5) such as:

- Gases like smoke, chemical compounds such as ammonia ($NH_3$), sulfur oxides ($SO_x$), nitrogen oxides ($NO_x$), carbon oxides ($CO_x$), volatile organic compounds (VOCs) and odors. The major greenhouse gas emissions from the waste sector are released to the atmosphere by the burning of solid waste materials, fossil fuels and wood products.

**FIGURE 2.5**

Air pollutants from waste incineration (Y. Divino, The Technology nowadays, http://moniegold.blogspot.it/2012/07/what-can-happen-if-you-dispose-of-your.html, available on line at 26/6/2014).

- Particulate matter (PM) such as $PM_{10}$ (10 micrometers in diameter or smaller), metals such as arsenic, beryllium, cadmium, chromium, lead (Pb), mercury and nickel.

Waste impacts caused by the abovementioned compounds could be summarized as follows:

- impact on human health such as disease transmission and public and property injury;
- impact on socio-economic conditions such as discouraging tourism and business;
- impact on environment such as causing damages to the ecosystem i.e. ground water, marine and coastal contamination;
- impact on climate such as greenhouse gases (GHG) emissions.

Environmental protection authorities and agencies in the world monitors waste impacts by issuing acts, laws, regulations and standards. In the following chapter, we will discuss the national and international laws related to waste treatment.

## Review Questions

What are the different types of waste impact?
Describe the impact of hazardous waste on human's life.
Explain the impacts of radioactive waste.
Describe the occupational impacts of waste.

## References

1. M. Vrijheid, 2000, "Health Effects of Residence Near Hazardous Waste Landfill Sites: A Review of Epidemiologic Literature", Environmental Health Perspectives, Vol. 108, Supplement 1, pp. 101–112.
2. D. Vallero, 2014, "Fundamentals of Air Pollution", 5th Edition, Academic Press.
3. Health impacts of solid waste, http://edugreen.teri.res.in/explore/solwaste/ health.htm, Adapted from UNEP report, 1996, available on line at 26/6/2014.

# 3

## Waste Related Laws, Regulations and Standards

*Key Learning Objective*

* Understanding the definitions and role of environmental laws, regulations and standards.

### 3.1 Introduction

This chapter provides an overview of international and national environmental regulations and laws related to waste management. In fact, most countries in the world protect their population and territories by issuing laws and legislations, out of these are the environmental laws that are related to waste management. An environmental law is a legal guideline that protects an environment by preserving the quality of air, water and land.

In the United States, there are various regulatory agencies that prepare regulations for environmental protection. Of these, the Environmental Protection Agency (EPA) offers resources for environmental legislations at the federal level [1,2]. The Occupational Safety & Health Administration (OSHA) and the National Institute for Occupational Safety and Health (NIOSH) provide regulations and standards in the fields of safety, health and environment [3,4].

In the European Union, the European Environment Agency (EEA) requires all 28 EU Member States to submit a report, under Directive 2001/81/EC, at the end of year that provides information on emissions of significant air pollutants [5]. In addition, the European Committee for Standardization, European Environment Agency (EEA) and UK Environment Agency offer several services in safety, health and environment [6].

In the Kingdom of Saudi Arabia (KSA), the Presidency of Meteorology and Environment (PME) provides the standards and regulations that aim to protect public health and maintain the preservation of the environment [7]. These regulations include the General Administration of Environmental Standards and General Environmental Law and Rules.

## 3.2  Basic Definitions

To better understand the environmental legislations, it is fundamental to distinguish between various terms used in waste legislations. The basic definitions of the terms *law, regulation* and *standards* are provided below:

- *Law* is *"a rule or set of rules by which a country is governed"* [8]. It is suggested by the government, passed by the parliament and signed by the president of the country.
- The definition of *regulation* by Dictionary of Environment & Ecology is *"the control of a process or activity"* [9].
- A *standard* is defined as *"something which has been agreed on and is used to measure other things by"* [9]. It is developed by non-governmental consensus committees.

An environmental legislation creates limits to duties and imposes a responsibility to produce outcomes in different places such as industrial zones (e.g. pollution impact). It is implemented by a government authority. In this sense, there are numerous legislations related to the reduction of waste and contaminants.

## 3.3  Environmental Legislations in Selected Countries

Selected international and national legislations related to waste processing and management are reported in the following paragraphs.

### 3.3.1  The USA

In the US, there are federal, states and local governments' legislations and regulations related to waste management. These are made available by several sources of information that can help individuals and companies decipher their obligations:

- *Federal or state agencies:* Agencies can be useful contacts to determine what statutes or regulations may be applicable.
- *Corporate resources:* Individuals that work for corporations can discuss environmental obligations with the corporate environmental health and safety department or the legal counsel. Corporations may even have an environmental compliance manual to assist in identifying obligations.
- *Federal Register:* Anyone can access regulations that are proposed and adopted by federal agencies through the Federal Register. The Federal Register can be accessed online at: www.nara.gov/fedreg. Proposed regulations contain information on where to send public comments and the deadline for submitting those comments. Public comments and background documentation for agency actions are also typically kept in dockets that are referenced in the Federal Register notices.

Waste management laws and legislations in the US deal with several targets that include products, pollutants and land uses. Legislations that target products deal with chemical substances and pesticide chemicals. Pollutants legislations regulate pollutants produced by industrial or domestic activities. The land use decisions are regulated at the local level. The Clean Water Act, for example, imposes obtaining permits before conducting dredge and fill operations.

Some US state laws implement federal programs in certain cases, such as:

- a state may implement federal programs established under the Clean Air Act or Clean Water Act if the state: (a) designates a lead state environmental agency; (b) adequately staffs, funds and designs the program to meet minimum federal standards;
- the state may set standards that are more stringent than federal standards;
- if the state program undermines the success of the federal program, the EPA may regain exclusive control over the program and administer the program itself.

Some state laws are independent from federal programs, such as:

- Massachusetts Toxic Waste Minimization Law that imposes mandatory waste reduction objectives on companies that use or generate toxic or hazardous wastes;
- California Proposition 65 which is an environmental-full disclosure law that requires companies to undergo significant efforts to make the public aware of health risks associated with products or environments to which they are exposed;

- New Jersey Property Transfer Environmental Law that requires extensive investigation and cleanup of contaminated sites before they are
  sold or transferred;
- Montana Solid Waste Laws—The Montana Solid Waste Management
  Act prohibits the disposal of any solid waste in any location not
  licensed as a solid waste disposal site by the Montana Department of
  Environmental Quality (DEQ).

The waste produced by industrial chemical and particulates processes and
their possible introduction into air, water or soil, such as toxic substances, is
regulated by Toxic Substance Control Act (TSCA) of 1976, administered by
EPA, that regulates over 75,000 chemicals produced or imported into the US.

Other agencies that play a significant role in environmental protection
established their own rules such as the Rules 59:43268–43280 (1994, August
22) issued by the Federal Registers for Hazardous Waste Operations and
Emergency Response. Furthermore, the OSHA regulates hazardous waste
operations and emergency responses by General Industry Standards [10].

### 3.3.2 European Union (EU)

In the EU countries, waste management policies are regulated by several directives such as Waste Framework Directive 2008/98/EC, Hazardous Waste
Directive 91/689/EEC, the Waste Oils Directive 75/439/EEC, and list of
establishing wastes, Decision 2000/532/EC, including a distinction between
hazardous and nonhazardous wastes. These directives make an obligation for
each member state to handle waste materials and their disposal safely [11].

### 3.3.3 Kingdom of Saudi Arabia (KSA)

Environment and human health protection against pollution risks in the
Kingdom of Saudi Arabia is regulated by the Presidency of Meteorology and
Environment (PME) regulations (www.pme.gov.sa/en/en_envrprot.asp).
The protection process deals with wastes and substances of all forms, and
independent of their sources. It also takes in consideration all stages of production, import, handling, storage, treatment and final disposal. This is in
addition to banning the import, entry or transit of hazardous waste materials
into the Kingdom.

The PME operates in accordance with the "General Environmental law
and Rules for Implementation and its Executive Regulation" which conforms
the Basel Convention that has been established in 1989 for controlling transboundary movement of hazardous waste and its disposal.

Article 14 of Chapter 2 in the General Environmental law and Rules for
Implementation identifies the agencies that are authorized to grant licenses
for waste handling and safe management. Article 11 seeks to enforce that

the design and operation of any project or activity complies with the applicable regulations and standards. Article 12 regulates those who work in the fields of treatment and disposal of solid wastes and implies that necessary precautions for safe storage and transportation of any waste types (e.g. gases, vapors, solid or liquid residues) are taken.

Environmental quality standards limits for air, water and land pollution, including guidelines for air pollutants, physiochemical pollutants, organic pollutants, inorganic pollutants, biological pollutants, are reported in the Environmental Protection Standards (EPS). Appendix-2 of the EPS reports the standards for environmental impact assessment of industrial and development projects. In addition, Appendix-3 is a guide for environmental accreditation procedures. The applicable hazardous waste control rules and procedures are provided in Appendix-4.

### 3.3.4 Other Arab Countries

Waste management in several other Arab countries suffers from poor environmental legislations, quantitatively, and poor implementation, qualitatively. Furthermore, in few Arab countries, foreign rules and regulations were enacted without any customization to suit the characteristics of the country [12], while others have their own proper laws and deal with waste management as a high priority issue.

For further information on environmental protection legislations, the reader is advised to consult the following websites:

- Occupational Safety and Health Administration (OSHA) Law & Regulations (United States):

  www.osha.gov/law-regs.html

  www.osha.gov/pls/oshaweb/owasrch.search_form?p_doc_type=standards&p_toc_level=0

- Environmental Protection Agency (United States):

  www.epa.gov/

- National Institute for Occupational Safety and Health (NIOSH) (United States)

  http://niosh-erc.org/

- The European Committee for Standardization:

  www.cen.eu/cen/AboutUs/Pages/default.aspx

  www.environment-agency.gov.uk/business/regulation/31865.aspx

- Presidency of Meteorology and Environment (KSA):
  www.pme.gov.sa/en/en_envrprot.asp

## Review Questions

Define the meaning of the terms: laws, regulations and standards.
What is the guideline of a waste management regulation?

## References

1.  Federal Laws, in "Laws, Regulations, and Guidance" at USEPA, available online at www.epa.gov/rpdweb00/mixed-waste/regs.html#laws, on 28/6/2014.
2.  Summary of Hazardous Waste Regulations, available online at www.dep.state.fl.us/waste/categories/hazardous/pages/laws.htm
3.  OSHA Law & Regulations, available online at www.osha.gov/law-regs.html
4.  NIOSH, available online at http://niosh-erc.org/.
5.  European Environment Agency, 2014, NEC Directive status report 2013, EEA Technical report No 10/2014, ISSN 1725–2237.
6.  European Committee for Standardization, available online at www.cen.eu/about/Pages/default.aspx; UK Environment Agency, www.gov.uk/government/organisations/environment-agency.
7.  PME in Kingdom of Saudi Arabia, available online at www.pme.gov.sa/en/en_envrprot.asp.
8.  P.H. Collin, Dictionary of Environment & Ecology, fifth edition, 2004, Bloomsbury, eISBN-13: 978-1-4081-0222-0
9.  Collin, Dictionary of Environment & Ecology.
10. Collin, Dictionary of Environment & Ecology.
11. OSHA, www.osha.gov/SLTC/hazardouswaste/index.html, available online at 13/7/2014.
12. European Union, http://ec.europa.eu/environment/waste/legislation/a.htm, available online at 19/7/2014.
13. www.waste-management-world.com/articles/print/volume-12/issue-4/temp/arabian-blights.html.

# 4

## Classification of Waste Materials

**Key Learning Objectives**

- Understanding the basis of waste classification.
- Understanding waste sources.

### 4.1 Introduction

Human activities are accompanied by the production of wastes as byproducts. Several factors increase the rate of waste production; these include the increase in population and the huge technological developments achieved in the last few decades. In fact, more complex waste substances are now produced that requires special management techniques in order to reduce their dangerous impact on human health and environment.

The accumulation of waste materials due to improper disposal is a major issue that causes problems like increasing the numbers of insect vectors like flies and mosquitoes, and scavengers such as stray dogs, pigs and rats which spread dangerous diseases. The accumulation of waste materials also generates bad odors and causes pollution. For example, at municipal level, the increase in population and urbanization are largely responsible for the increase in generation of waste especially in big cities. If every person living in a city that has a population of one million generates 1 kilogram of waste daily, around 1,000 tons of waste will be produced in the city every day; this reflects the severity of the issue world-wide.

For a better understanding of the severity of the problem and to work toward a solution, it is necessary to understand the types of waste that are being generated. Furthermore, to provide an accurate data on this problem, it is necessary to understand the quality and quantity of the waste materials, their nature and constituents.

## 4.2 Classification of Waste Materials

Waste materials are complicated matters and their classification process is complex. In fact, they are very heterogeneous and have a great variation in composition since they come from different sources and different locations. However, the classification is necessary for the consistency in the description of waste, its management and disposal.

There are several terms in use in the classification of wastes. The wastes can be classified by theirs sources, materials states, management or framework and by their properties.

### 4.2.1 Based on the State

Waste materials can be classified on the basis of their state into solid, liquid and gas:

*Solid waste*: waste in solid state include

- municipal and domestic waste materials that include materials such as construction materials, plastics containers, different types of bottles and cans, papers and scraps;
- industrial solid wastes that include hazardous and nonhazardous wastes, clinical and hospital wastes, household hazardous wastes and sewage sludge;
- any other solid wastes such as agricultural wastes, special wastes, such as scrap tires and end-of-life vehicles, power stations ash, mines and quarry wastes.

*Liquid waste*: Liquid waste materials include liquid discharges from industrial plants and domestic washings and several other sources.

*Gas waste*: Industrial wastes such as chemical gases wastes from manufacturing industries and other sources such as gaseous pollutants, odors, suspended particle matter (dust, fumes, mist and smoke).

*Radioactive waste*: It is defined as any radioactive waste material, solid, semisolid or liquid that contains radioactive elements. The main sources of these wastes are industrial activities and weapons.

### 4.2.2 Based on the Source

There are different human activities that produce waste, the most common sources are:

*Domestic waste*: This is mainly generated by residential activities that include waste materials generated from houses, such as paper, plastic, glass, vegetable waste.

*Commercial waste*: Waste materials generated by commercial and business sectors which includes waste generated by commercial establishments or wholesale, like shops, spoiled goods, vegetable, retail, restaurants and meat remnants, yard trimmings, office buildings or office waste materials such as printer papers, disposable tableware, paper napkins.

*Institutional waste*: This type of waste materials is produced by public institutions such as educational institutions and others.

*Agricultural waste*: This generally refers to solid waste that is generated from crop production and harvesting or rearing of animals, like crop discards (leaves) from livestock or trees, vines, twigs, branches and weeds. It also includes animal wastes (manure) produced from farms, like animal dung, used bedding and carcasses.

*Construction waste*: This type of waste includes all materials produced by construction processes such as demolition and all types of construction and natural disasters remnants. It also includes metal or steel rods, steel girders, wood, bricks, cement, concrete, window glass, wiring, drywall, plaster, insulation, asphalt, lumber, roofing materials, plastic piping, etc. and other miscellaneous items related to these activities.

*Industrial waste*: It is the type of waste generated by manufacturing processes. All industries generate some waste such as excess materials from manufacturing which can be divided into nonhazardous waste or hazardous waste. For example, garment factory would dump textiles of various kinds, pulp and paper, iron and steel, glass, plastics and concrete. Waste from extraction industries includes waste from mining and mineral processing, metals, minerals, acids and solvents.

*Energy production (fuel waste such as ashes)*: This waste comes from the burning of solid fossil fuels like coal, wood and coke or, sometimes, from open waste burning, oil and gas industry waste, including solids and liquids produced in exploration, drilling and production of crude oil or natural gas.

*Medical waste*: This refers to waste generated from the care houses and hospitals during clinical and research activities. Medical waste materials are considered very infectious and include expired or unused drugs or medicines, anatomical waste such as body fluids, cultures, needles, swabs, bottles and tubing, plastic, syringes, surgical dressings, bandages, wraps, bedding, medical and dental devices, and protective clothing and chemical wastes.

*Waste from wastewater treatment stations such as sludge*: This refers to sewage, household or industrial wastewater discharged into sewers, dredging, waste solids and semisolids removed from the bottom of rivers and harbors.

### 4.2.3 Based on Biological Properties

According to their biological properties, waste materials can be divided into biodegradable and non-biodegradable waste.

*Biodegradable*: All organic waste materials such as vegetables, fruits, flowers and kitchen waste materials can be degraded by biological processes. The biodegradable waste materials can be decomposed by the natural processes and converted into their elemental forms. For example, all biomass materials such as papers, wood, fruits, kitchen waste and animal dung are biodegradable.

*Non-biodegradable*: These include materials that cannot be decomposed and remain as is in the environment for long periods of time. They are persistent and can cause various environmental problems.

### 4.2.4 Based on Waste Hazardous Properties

The impact of waste materials depends on their composition and properties. Infectious waste materials are medical waste that contains pathogenic microorganisms that can cause diseases. However, medical waste materials can be infectious or noninfectious; it is estimated that more than 25% of hospital waste materials are infectious (Muhwezi, 2014 [1]).

Based on their impact on society and environment damaging, waste materials can be divided into two main categories: nonhazardous and hazardous waste. Nonhazardous waste materials refer to substances that are safe to be treated and do not have serious impacts on human health and the environment.

Hazardous waste materials refer to substances or any unwanted or discarded materials that are unsafe to be dealt with because of their physical, chemical or infectious characteristics that could cause significant hazard to human health or the environment. By law and scientific manners, these materials must be monitored in terms of improper treatment, storing, transportation or disposal. Hazardous waste materials include corrosive, highly inflammable, toxic and explosive substances. In addition, certain types of household waste materials make part of hazardous materials (e.g. solvents, pesticides and products containing volatile chemicals.)

Toxic waste materials refer to those materials that are poisonous in nature. These include chemicals, medicines, paints, bulbs, tube lights, fertilizer, tires, pesticides, batteries and cans. In fact, toxic materials create health problems for the people living close to areas that represent sources of these materials. Among other health problems that are caused by such materials, we can mention nausea, allergies and eyes irritation.

The main sources of industrial waste materials include metal industries, refining processes and chemical industries. Mercury and cyanide are examples of hazardous materials that are at direct exposure at industrial sector and they can be fatal.

Special hazardous waste materials include radioactive, explosives and electronic materials (e-waste). The radioactive waste materials are waste substances that consist of unstable isotopes and particularly dangerous because they cause several lasting damages to human beings. In fact, over time, these materials decay to a more stable element or form emitting potentially harmful energy in the process of transformation as well as the mass–energy relationship. The transformation is accomplished by several different mechanisms, most importantly alpha particle, beta particle and gamma ray emissions; these have impacts such as creating changes in the genetic structure of individuals (called mutation).

Sources of radioactive waste include nuclear power plants, medical radiotherapy devices in hospitals, mining waste such as phosphate mines, nuclear defense or military government waste, normally occurring radioactive materials and nuclear or radiological accidents. Also, radioactive wastes are present in several forms or concentrations.

### 4.2.5 Based on Framework

Based on the framework, waste materials can be divided into controlled and non-controlled waste. Non-controlled waste materials include radioactive, explosive and agricultural materials.

Controlled waste materials include household or domestic waste, commercial waste, medical or clinical waste and industrial waste. These, in turn, could also be divided into nonhazardous waste, hazardous waste or special waste.

An example for classification of waste based on framework is the management of municipal solid waste (MSW). It includes solid waste materials produced in urban areas. This comprises miscellaneous, or mixed, materials that are not sorted into specific categories. Typical examples of MSW are household waste materials, commercial waste materials, institutional centers and construction and demolition debris.

Furthermore, waste materials can be divided into recyclable and non-recyclable waste. Recyclable wastes are materials that can be recovered for recycling purposes. The recycled materials are either converted into raw materials or used in producing new products. On the other hand, non-recyclable materials are treated and disposed of.

For further information, the reader is advised to consult the following:

http://cmsdu.org/organs/Solid_Waste_Management.pdf on 24/4/2016.

http://wonder.cdc.gov/wonder/prevguid/p0000019/p0000019.asp on 24/4/2016.

https://en.wikipedia.org/wiki/Solid_waste_policy_in_the_United_States on 24/4/2016.

www.southernsolidwaste.com/waste-terminology/ on 24/4/2016.

www.ehso.com/Glossary.htm on 24/4/2016.

http://edugreen.teri.res.in/explore/solwaste/types.htm on 24/4/2016.

www.truhealthonline.com/ on 24/4/2016.

## Review Questions

Mention the bases of waste classification.

How do you classify the waste produced by your community?

Describe the health impacts caused by hazardous wastes.

What is a radioactive waste and what are the different categories of radioactive wastes?

## Reference

1.  L. Muhwezi, P. Kaweesa, F. Kiberu and E.L.I. Eyoku, 2014, "Health Care Waste Management in Uganda: A Case Study of Soroti Regional Referral Hospital", International Journal of Waste Management and Technology, Vol. 2, Supplement 2, pp. 1–12.

# 5

## Waste Management Hierarchy

*Key Learning Objectives*

- Understanding the terminology of waste management.
- Understanding the terminology of waste hierarchy.

## 5.1 Introduction

Waste materials play an important role in our lives; more specifically, these materials could have harmful effects on our health and the environment. In Chapter 4, we have classified wastes into liquid wastes, solid wastes and gas wastes (air pollutants). Often, these three categories are interrelated, for example, air pollutants can be removed from an air discharged by a scrubber, and the waste scrubber solution could contaminate the water if it is not well handled according to water quality regulations. In addition, organic solid waste produces leachate that enters the groundwater and transfers pollutants into the water body.

Therefore, efficient processing and management technologies are deemed necessary to get rid of waste materials in the most efficient way. The technologies of waste treatment are associated with several kinds of products and processes.

This chapter describes the main methods of management applied to all waste types; it also explains how the waste materials are minimized in manners that are compatible with the engineering, health, safety and environmental regulations and laws.

## 5.2 Waste Management Strategies

The main objective of a management strategy is reducing the amount of waste sent to final disposal. The complete elimination of waste materials seems to be impossible; however, it is possible to reduce its sources and generation. The strategy that aims to reduce the waste to the minimum, wherever possible, is known as zero waste strategy.

Waste management methods and treatment processes depend on the manufacturing processes, plant size and daily waste flow rates. Waste treatment processes can transfer substances from one of the three waste categories (solid, liquid and gas) into any of the others.

The word *technology* is defined in the "Dictionary of environment and ecology" as "the application of scientific knowledge to industrial processes [1]". Also, the word *treatment* is defined as the use of a chemical, physical or biological process to something in order to get a specific result. In fact, treatment involves modifying a waste's physical, chemical or biological character or composition through designed techniques or processes.

Waste treatment technologies include several large processes (see Chapter 7) that depend in their application on waste quantity and characteristics (including hazardous or nonhazardous waste). Waste material can be managed in three steps:

1. Analyzing waste material source
2. Analyzing the available processes and treatment technologies
3. Minimizing and reduction of waste material production. There are three methods of waste management and minimization, these are source reduction, recycling and treatment. Here is a brief definition of each:

    *Source reduction*: Any procedure applied to reduce waste materials at the point of generation.

    *Waste recycling*: Waste reduction by sorting out recyclable components after they have been mixed together.

    *Waste treatment*: Reducing the amount of waste after production and collection.

Process modification of an existing plant is a form of source reduction method; however, the best approach takes the waste minimization process in consideration in the phase of plant designing.

## 5.3 Waste Management Hierarchy

It has been mentioned in Chapter 3 that local laws and regulations in each country govern the waste treatment and management. Waste treatment refers to the activities required to ensure that waste has the least practicable impact on the health and environment, which includes the production, management and disposal of waste; this is often known as *"waste hierarchy"* (Figure 5.1). The waste hierarchy is a nationally and internationally accepted guide for prioritizing waste management practices with the objective of achieving optimal health and environmental outcomes. This is mainly based on strategic policy set by the individual countries.

The hierarchy provides a classification of waste management according to the strategy of the individual country. It is composed of a series of five or six options to implement a waste management plan. The most favored option is the prevention and the least favored option is the disposal of. The different forms of waste treatment are graded in the waste hierarchy. Table 5.1 shows the summary of the waste hierarchy; each section of the hierarchy has specific area of waste management. Facilities are needed to implement waste management that will be required to sort, reduce, recycle and process waste before final disposal. It is also necessary to set in place effective policies within development plans and projects.

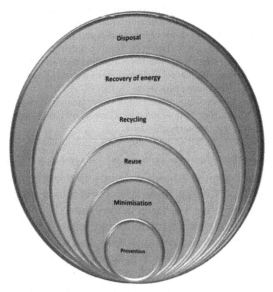

**FIGURE 5.1**

Hierarchy of waste management.

**TABLE 5.1**

Brief description of the waste management hierarchy

| Section | Waste Management |
| --- | --- |
| Prevention | It is the highest point on the hierarchy and the most preferable strategy, starting from industrial design that takes into consideration the strategy of zero waste production. This means thinking about waste prevention and not treatment of waste, before and after production. An example on the case before production: thinking about no waste production; how to increase process efficiency and resources in phase of design? how to eliminate the unnecessary usage of raw materials? using new raw materials with minimum scattering and/or how to keep the products for longer time. After production completion, recovery and reuse of materials through their recycling, wherever possible, could be implemented.<br>This section of hierarchy requires increasing the public awareness on the zero waste or waste minimization concept and additional governmental efforts to support this approach. |
| Minimization or reduction | Waste minimization or waste reduction is a policy for reducing the amount of waste that is produced by a society (company or community). This is usually measured in kg produced per person per day. The enforcement of this concept requires adequate legislations and laws. |
| Waste recovery or preparing for reuse | This is considered as one of the techniques used to realize the minimization or reduction concept. It is a strategy of separating and sorting from mixed waste the different categories or types of waste with the aim of recovering of materials to be reused or recycled.<br>As an example, this section is applicable in the practical treatment of municipal solid waste (see Chapter 6). |
| Recycling | The recycling is a process of transforming materials into new raw materials (secondary resources) or reusing. |
| Energy recovery | It is a strategy to reduce the waste and recovery of energy from it by biological (anaerobic digestion) or thermal (incineration, gasification and pyrolysis) processes. |
| Disposal | This phase comes at the end of waste treatment process. It represents the proper disposition treated waste materials according to the local environmental regulations. |

In the EU, waste hierarchy has five steps that are regulated by Waste Framework Directive 2008 [2]; these are: prevention, preparing for reuse, recycling, other recovery (e.g. energy recovery) and disposal. There are many strategies in waste management for reducing waste which are called Rs as Reduce, Reuse and Recycle. In certain strategies, there are 3 Rs while there are 4Rs or more in other strategies. An example of 4Rs strategy is Reduce, Reuse, Recycle and Refuse and a 5Rs strategy include Reduce, Reuse, Recycle, Refuse and Recover.

These Rs are meant to be a hierarchy, in order of importance, which classify waste management strategies according to their desirability and policy

**TABLE 5.2**

Description of Rs strategies in waste management

| Strategy | Description |
| --- | --- |
| Reduce | A strategy for reducing the generation of unnecessary waste mass using techniques such as recycling or reusing, e.g. to reduce the plastic bags: when you go to the market carry your own shopping bag made of cloth or jute and put all of your purchases directly into it. |
| Reuse | A strategy of using waste materials that could be reused for similar or different purpose, e.g. reuse the plastic bags for shopping again and again. |
| Recycle | A strategy that has the main purpose to sort and separate waste materials to recover the recyclable materials or to create new material or products, e.g. recycle the organic wastes by composting (see Chapter 8) to produce a humus (fertilizer for soil amendment). |
| Refuse | A strategy for the reduction of waste mass refusing to buy new items though you may think they are prettier than the ones you already have, e.g. instead of buying new containers from the market, use the ones that you have in house. |
| Recover | Strategy for recovering materials and/or energy from waste; it is entering in Reduce, Reuse and Recycle of wastes strategies. For example, the recovery of energy (biogas) from organic wastes by anaerobic digestion (AD). |

of authority. On the basis of this definition, waste hierarchy refers to the Rs management strategies. For example, the strategy is not how to prevent producing waste, but how to reduce it. Reusing is usually followed by recovery techniques, while disposal comes at the bottom of the hierarchy; it represents the worst option. Table 5.2 provides a brief description of the Rs.

The advantages of Rs strategies or waste minimization are:

- reducing the toxicity of waste;
- reducing greenhouse gases that cause global warming;
- saving costs such as transportation and disposal costs;
- saving non-renewable energy resources;
- reducing energy consumption;
- reducing quantity materials that are moved to landfills;
- reducing quantity materials that are moved for incinerator usage;
- reducing the problems associated with waste treatment or disposal.

Successful implementation of management strategies of waste reduction at the practical level requires setting up an appropriate national plan that disseminates the concepts of waste minimization and increases the public awareness on the importance of these strategies. This might be achieved by conducting training courses about the importance of waste minimization strategy to industrial production managers and for students and civil organizations. Furthermore, at the industrial level, training courses and workshops can be organized to educate workers about this strategy. Training could include directions on the use of new technologies or modified equipment or production process or operating conditions that lead to reduction in waste production at the source. This might also include the use of recycling and recovery techniques or the reuse of the waste materials. In the next chapters we will discuss the processes of waste treatment and technologies of waste minimization.

## Review Questions

What is waste management?
What are the differences between waste disposal and waste minimization?
What is meant by waste recycling?
Describe the waste hierarchy.

## References

1. P.H. Collin, Dictionary of environment and ecology, Fifth edition, 2004, Bloomsbury, eISBN-13: 978-1-4081-0222-0
2. Sepa, available online at www.sepa.org.uk/waste/moving_towards_zero_waste/waste_hierarchy.aspx, on 14/6/2014.

# 6

## Solid Waste Management

*Key Learning Objectives*

- Understanding the concept of solid waste management.
- Understanding the concept of waste disposal.

### 6.1 Introduction

Increase in the population, urbanization and prosperity led to a rapid increase in the production of solid waste materials. Thus, solid waste management becomes a major challenge to municipal authorities in the collection, transport, recycle, treatment or disposal phases. As for example, in the US, the estimated annual industrial solid waste generation and disposal is about 6.6 billion [1]. Waste is being generated at every stage of industrial production process and it can be classified into two main categories: hazardous waste and nonhazardous waste.

Waste treatment and processing imply the use of multiple technologies whose application depends on local, regional and national capabilities that include regulations, policies, economic priorities and practical local limits (infrastructure).

Hazardous waste consists of more than a single chemical or substance that has specific characteristics such as ignitability, corrosively, reactivity and toxicity. On the other hand, nonhazardous waste is composed of mining and mineral processing materials, coal ash, cement kiln dust, foundry sands, oil and gas production waste and other waste materials that lack the characteristics of hazardous waste.

Most industrial waste comes from important industries such as petrochemicals, chemical and pharmaceutical industries, fertilizer and agricultural chemical production, plastic and resin manufacturing. Industrial waste also includes the metallurgical manufacturing, cement, stone, glass, clay and concrete production industries. Textile and leather manufacturing, rubber products manufacturing, electric power generation, pulp and paper manufacturing and food production and processing make part of industrial waste.

In this chapter, the focus will be on the treatment of both types of solid waste, namely, the industrial and municipal solid waste (MSW).

## 6.2 Solid Waste Management

The process of solid waste management is based on the origin and contents of waste materials; specifically, solid waste originating from domestic and commercial activities that could be organic or inorganic.

Waste management process starts with monitoring that is followed by collection, transport, processing, recycling and disposal. Furthermore, waste could be classified into two major classes: hazardous, i.e. toxic materials, and nonhazardous. The two classes are divided into the following five categories:

A) *Nonhazardous wastes*
   1. Municipal waste
   2. Nonhazardous industrial waste
   3. Agricultural waste
B) *Hazardous wastes*
   1. Hospital hazardous waste
   2. Industrial hazardous waste

The management and treatment of hazardous waste need a special treatment and procedure and is out of the scope of this handbook. For detailed information about the management of hazardous industrial waste, the reader can refer to the book "Advances in hazardous industrial waste treatment/edited by Lawrence K. Wang, Nazih K. Shammas, Yung-Tse Hung", 2009 [2].

## 6.3 Management of Nonhazardous Solid Waste

Treatment of nonhazardous industrial waste materials has several advantages; among these are: (a) reducing the volume and toxicity of waste prior to disposal; (b) making waste materials amenable for reuse or recycling; (c) physical treatment involves changing the waste materials' physical properties, such as its size, shape and density, this will facilitate the successive processes.

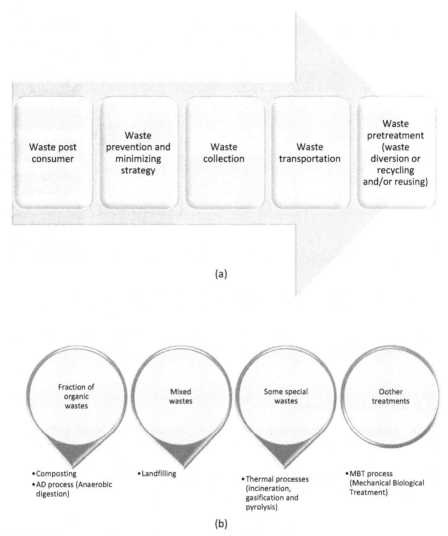

**FIGURE 6.1**

(a) Illustration of the steps of solid waste pretreatment and (b) treatment of major processes applied to organic solid waste.

Figure 6.1a illustrates the steps of the solid waste processes starting from waste post-consumer until waste pretreatment. Figure 6.1b shows the processes applied to organic solid waste. The first steps of treatment need low to intermediate technology and cost, while the last steps, such as thermal processes, Mechanical Biological Treatment (MBT) process and Anaerobic Digestion (AD) process need high technology and cost. Major solid waste disposal strategies include landfilling, composting and incineration. Figure 6.2

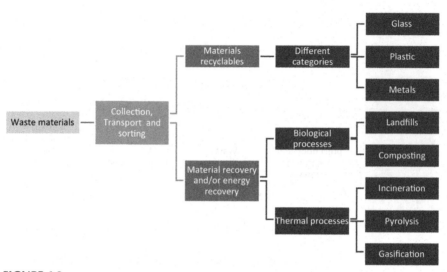

**FIGURE 6.2**

A diagram representation illustrating the steps of solid waste treatment processes.

shows the steps of processes for solid waste treatment with strategies of recycle and recovery (material and energy).

### 6.3.1 Monitoring Process

Waste management monitoring must be conducted by local authorities (municipal) through enforcement of law. Monitoring is subjected to the applicable regulations and appropriate control measures to eliminate or reduce adverse impacts of wastes on the human health, environment and to improve quality of life.

### 6.3.2 Generation and Collection Process

Waste materials are produced by the producer (residents and other generators) at any time of the day; these are generally put inside a shared container. Then, the collected waste materials will be removed by local authorities through regular waste collection and transport means. Certain regulations, in this case, must be taken to consideration because the waste storage containers are located inside residential areas. The most significant considerations include: (a) avoid odors that result from decomposition of organic matters in the accumulated waste materials; (b) shared containers must be protected and harbored from fly breeding and other pests.

**FIGURE 6.3**

The strategy of source separation of wastes at the point of generation. (a) One bin for mixed recyclables; (b) two bins for two types of separation; (c) three types of separation; (d) four types of separation; (e) five types of separation; (f) six types of separation.

Waste generation in large cities requires affordable, environmentally effective and sustainable waste management plans. In fact, these depend on the policy, planning and economical capabilities of local authorities. For example, in many countries the collection of urbane waste is made using bins posted in streets near residents, sometime there are special collections for recycling. In this case, there is a need to increase public participation and providing training and educational programs to increase public awareness on the importance of these processes. Figure 6.3 illustrates the strategy of waste materials source separation and keeping different categories of recyclables at the point of generation. The color codes of bins are used to distinguish the different waste types; for example, four colors of bins are used for the separation of plastic, glass, metal and paper materials.

Separation of waste materials at the point of generation helps and facilitates the recovery of materials and categorizes them into recyclable or reusable materials. Figure 6.4 illustrates different categories of collected materials for recyclables.

### 6.3.3 Transport Process

Transportation of waste materials is the movement of waste products from the generation location or first step of waste storage to the treatment or sorting location (second step of waste storage). This process depends on the quantity of waste materials to be transported. In addition, the collected waste quantity determines the most adequate mean of transport and equipment.

The rate of the waste materials generation is estimated using statistical methods over a period. It is measured by daily kilogram produced by one person (kg/capita/day). This rate has three levels: low, medium and high. Another practical method to estimate this rate is by weighing the vehicle and its load of wastes when it enters the treatment station then subtracting the tare weight of vehicle (weight of the empty vehicle).

|  |  |  |
|:---:|:---:|:---:|
| (a) | (b) | (c) |
| (d) | (e) | (f) |

**FIGURE 6.4**

Different categories of collected recyclables (a) plastic waste materials (Intercon, Recycling vs. Upcycling: What is the difference? available online at http://intercongreen.com/2010/ 02/17/recycling-vs-upcycling-what-is-the-difference/ on 13/06/2020); (b) aluminum waste materials (wisegeek, available online at www.wisegeek.com/what-is-the-importance-of-recycling.htm#comments on 13/06/2020); (c) glass waste materials (J. Fenston/WAMU, APR 26, 2019: Arlington Ends Curbside Glass Recycling, available online at https://wamu.org/ story/19/04/26/arlington-ends-curbside-glass-recycling/ on 13/06/2020); (d) batteries waste materials (Fresh ideas/womenworld, available online at http://womenworld.org/health/ fresh-ideas--you-can-make-a-difference-to-the-environment.aspx on 13/06/2020); (e) organic food waste materials; (f) electronic waste materials. Source: Intercon, it is in the public domain.

The transportation cost of waste materials depends also on the distance of movement and the types of transport vehicles. Far distances imply increased costs (see Chapter 12). In majority of countries, the residents contribute in the waste cost by annual contribution. The type of transportation used depends on the quantity of waste materials to be moved, distances and capabilities of local authorities. The means of transport could be divided into human powered, animal powered and motorized. Figure 6.5 illustrates these three means of solid waste transportation. A brief description of these means is provided below.

*Human powered:* Wheelbarrows are used for collection and transportation of waste from streets and markets; this is being pushed or pulled by a human being for short distances.

*Animal powered:* Animals like donkeys or horses are used to pull carts; these are used, often, in poor countries or rugged areas, and used for short-distance transportation.

  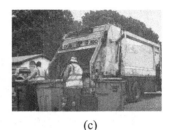

(a) (b) (c)

**FIGURE 6.5**

Solid waste material transportation means: (a) human powered (Pikist and Pxfuel, Free to use image, available online at www.pxfuel.com/en/free-photo-eoyzs, and www.pikist.com/free-photo-xfcgu on 18/06/2020); (b) animal powered (Castelbuono.org, available online at www.castelbuono.org/il-sindaco-scrive-una-lettera-aperta-alla-comunita-di-castelbuono-in-merito-alla-raccolta-dei-rifiuti/, https://www.castelbuono.org/wp-content/uploads/2019/09/1182014124147c.jpg on 18/06/2020); (c) motorized.

*Motorized vehicles*: Different types of vehicles are used for collection and transportation of solid wastes; these are used for different distances of transportation and volumes of waste materials. These vehicles are used for transportation in all phases starting from waste materials collection, storage, sorting and the ultimate disposal of these materials. The selection of vehicle type depends on the economic capabilities of the local authorities.

### 6.3.4 Waste Sorting

Waste sorting is the process of separating waste materials into different categories with the aim of reuse, recover or recycle waste materials. It is convenient to implement the sorting process at the point of generation. In the case that the waste materials separation at the point of generation strategy is applied, the sorting process can occur manually by the producer at the site of collection. If done at a sorting station, sorting process could be done either automatically and/or manually depending on the waste type and available facilities (Figure 6.6).

In the automatic or semiautomatic sorting, waste materials are sent along a conveyor belt with a series of sensors (Figure 6.6b); for different types of waste materials (in case of sorting automatically), there are different separation technologies that are based on (a) waste density (as X-ray technology); (b) reflected light (as near infrared, NIR, technology); (c) separation by a system of fast air jets; (d) separation by an electromagnetic system for metal separation; (e) drum screen or trommel for separation according to waste materials particle size. In certain cases, human assistance is needed in the case of semiautomatic systems at the conveyor belt; this is called manual picking line.

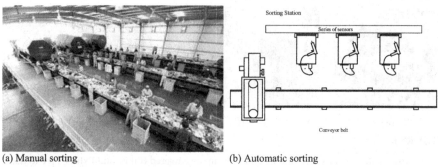

(a) Manual sorting                                (b) Automatic sorting

**FIGURE 6.6**

Waste sorting types at a sorting station. (a) Manual MSW sorting (recycling plant—Saudi Arabia) (Sweep-net.org, Manually MSW sorting (recycling plant—Saudi Arabia), available online at www.sweep-net.org/organic-waste-business on 14/6/2014); (b) automatic MSW sorting station.

The sorted waste materials can be divided into two main categories, namely materials that could be recycled and those that cannot be recycled. The recycled materials, such as glass, aluminum, rubber and paper, could be the raw materials for several industrial processes, as for example, melting glass to produce new products from the recycled material. A fraction of organic waste is usually destined into composting or anaerobic digestion process. This type of waste materials could be recycled naturally in different methods. This takes place firstly by composting process to be used as fertilization; secondly by anaerobic digestion to recover energy as biogas; and lastly by combustion process.

The separation of organic waste process is called Mechanical Biological Treatment (MBT), and the plant used in this process is called MBT plant (see Chapter 8).

Following the above-mentioned processes, the waste materials are destined into final or ultimate disposal; this includes composting, anaerobic digestion (AD), landfill and incineration.

## Review Questions

What is the problem of industrial solid waste?
What is meant by MTB?
Explain the factors that control the waste transport process?

## References

1. EPA's Guide for Industrial Waste Management: Introduction, available online at www.epa.gov/epawaste/nonhaz/industrial/guide/index.htm on 14/6/2014.
2. L.K. Wang, N.K. Shammas and Y.-T. Hung (eds.), 2009, "Advances in hazardous industrial waste treatment", CRC Press, Taylor & Francis Group, ISBN-13: 978-1-4200-7230-3.

# 7

# *Waste Treatment Processes*

*Key Learning Objectives*

- Understanding the terms waste processes and waste disposal.
- Understanding the processes of waste treatment.

## 7.1 Introduction

As mentioned in Chapter 5, the methods of waste management, reduction and minimization at sources or disposal of depend on the type of applied processes. Engineering technologies include several processes such as physical, chemical, physio-chemical and biological or microbiological technologies [1]. In this chapter, we discuss the main processes of waste treatment and disposal.

## 7.2 Processes of Waste Treatment and Disposal

In several engineering technologies, waste treatment processes operate in a complementary and integrated manner such as Mechanical Biological Treatment (MBT) processes. Table 7.1 shows the advantages and disadvantages of these technologies. Selecting the type of process depends on the type of waste and its characteristics.

**TABLE 7.1**

Advantages and disadvantages of different environmental engineering technologies of waste treatment

| Type of technology | Advantages | Disadvantages |
|---|---|---|
| Physical technologies (sedimentation, filtration, volatilization, fixation, evaporation, heat treatment, radiation, etc.) | Required time is from few seconds to few minutes; high predictability of the system | Low specificity and high energy demand |
| Chemical technologies (oxidation, incineration, reduction, chemical immobilization, chelating, chemical transformation) | Required time is from few seconds to few minutes; high predictability of the system | High expenses for reagents, energy and equipment; air pollution due to incineration, formation of secondary wastes |
| Physio-chemical treatment (adsorption, absorption, chromatography) | Required time is from few minutes to few hours | High expenses for adsorbents; formation of secondary waste |
| Microbiological technologies (biooxidation, biotransformation, biodegradation) | Low volume or absence of secondary hazardous wastes; process can be initiated by natural microorganisms or small quantity of added microbial biomass; high process specificity; wide spectrum of degradable substances and diverse methods of biodegradation | High expenses for aeration, nutrients and maintenance of optimal conditions; required time is from some hours to days; unexpected or negative effects of microorganism's destructors; low predictability of the system because of complexity and high sensitivity of biological systems |

Source: Volodymyr Ivanov, 2010.
A description of the processes of waste treatment are described in the following sections.

## 7.2.1 Physical Processes

A physical process does not modify the waste chemical composition, but it changes its physical properties such as state (i.e., gas, liquid, solid), size, shape and density. Examples of physical processes include grinding, shredding, compacting, screening, sedimentation, filtration, evaporation/volatilization, immobilization (e.g. encapsulation and thermoplastic binding), carbon absorption (e.g. granular activated carbon and powdered activated carbon), solidification/addition of absorbent material, distillation (e.g. batch distillation, fractionation, thin film extraction, steam stripping and thermal drying) and so on.

## 7.2.2 Chemical Processes

A chemical process involves altering a waste chemical composition, structure and properties through chemical reactions. The process is based on mixing the waste material with other reagent material and/or heating the waste at high temperatures. Through these chemical modifications, the waste constituents can be recovered or destroyed. Some examples of chemical processes are stabilization, neutralization, vitrification, oxidation, reduction, precipitation, chlorination, acid leaching, extraction (solvent extraction and critical extraction), ion exchange, incineration, thermal desorption, high temperature metal recovery and so on.

## 7.2.3 Biological Processes

The biological processes are explained with some details in the following paragraphs while their applications are discussed in Chapters 8 and 10.

Biological processes can be divided into two types: aerobic and anaerobic. The aerobic (presence of oxygen) process uses microorganisms to decompose organic matters into carbon dioxide, water, nitrates, sulfates, simpler organic products and cellular biomass (i.e. cellular growth and reproduction).

On the other hand, an anaerobic (absence of oxygen) process uses microorganisms to transform organic constituents into carbon dioxide and methane gas. Examples of biological processes include aerobic process of activated sludge, aerated lagoon, trickling filter and rotating biological contactor (RBC) and the anaerobic digestion (AD).

The biological process depends on the microorganisms that contribute to the biodegradation of biomass or organic materials or biowaste. The biodegradation of organic materials is chemically very complex; the breakdown of the large organic molecules into small compounds or elements needs the presence of a catalyst or an enzyme (provided by the microorganisms).

Biowaste is an organic matter that is composed of the elements carbon, hydrogen, oxygen, nitrogen, phosphorus and other elements in the form of compounds or complex organic molecules. It is used as food for the microorganisms during their own metabolic degradation process in their life. Then, microorganisms consume the biowaste to get energy for growing and to maintain themselves.

Microorganisms include several species of bacteria, fungi, algae, ciliates, rotifers, protozoa, nematodes and worms. Madigan et al. (2012) [2] provided detailed information about microorganisms and their biology. The biological activity depends on the internal and external environmental conditions that help the microbial growth for optimal operation of the process. The most important factors affecting the biological processes are temperature, pH, humidity, agitation, retention time and feedstock. Reineke (2001) provided detailed information about aerobic and anaerobic biodegradation processes [3].

## 7.3 Controlling Factors of Biological Processes

Biological processes that are used in waste treatment are controlled by several factors; these are discussed in detail below.

### 7.3.1 Microorganism Growth

There are different phases for microorganism growth in biological processes. In the bioreactor, the cultures of microorganism can be operated in batch (discontinuous), fed batch (semicontinuous) and continuous mode. In the batch process, plotting the logarithm of biomass (cell) concentration, X, against time, t, will produce a growth curve. Figure 7.1 shows the typical growth curve of a microorganism in batch culture after inoculation into a new growth medium.

The growth rate is defined as [4]:

$$\mu = 1/x(\mathrm{d}x/\mathrm{d}t) \tag{7.1}$$

Where:

$\mu$ = the specific growth rate ($\mathrm{h}^{-1}$);

$x$ = the biomass concentration of microorganism ($\mathrm{g\ L}^{-1}$);

$t$ = time (h);

$\mathrm{d}x/\mathrm{d}t$ = the variation in biomass concentration with respect to time ($\mathrm{g\ L}^{-1}\mathrm{h}^{-1}$) (culture productivity per unit volume), or the growth of concentration of organisms.

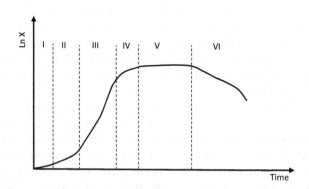

**FIGURE 7.1**

Typical growth curve of a microorganism in batch culture during the six phases.

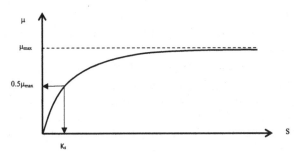

**FIGURE 7.2**

Relationship between the substrate concentration (S) and growth rate ($\mu$).

The specific growth, $\mu$, as in the original mathematical equation described by Monod [5], is a function of the concentration of a substrate, S, in the case of medium limitation concentration (Figure 7.2):

$$\mu = \mu_{max}\, S/[Ks + S] \qquad (7.2)$$

Where:

S = concentration of a substrate;

$K_s$ = a saturation constant;

$\mu_{max}$ = the maximum specific growth rate of the microorganism.

The rate of growth depends on the growth phases (Figure 7.1); these phases are (I) lag phase; (II) acceleration phase; (III) exponential phase; (IV) deceleration phase; (V) stationary phase and (VI) death phase. A brief description of each phase is provided below:

  I. *Lag phase*: A phase necessary for a microorganism to become accustomed to the new environmental conditions (substrate, temperature, pH and so on) for growing; the cells start to build their structural components or enzymes (synthesize) and prepare to grow. In this phase, there is a very little growth and the specific growth rate, $\mu$, is zero.

 II. *Acceleration phase*: It is the phase in which the microorganisms get used to the new habitat and they start to grow quickly; the specific growth rate in this phase is less than $\mu_{max}$.

III. *Exponential phase*: In this phase, the microorganism growth is increasing rapidly until the specific growth achieves its maximum rate, $\mu = \mu_{max}$. This phase can be extended by removing inhibitions or by adding new nutrients or both. In this case, the process becomes semicontinuous.

IV. *Deceleration phase*: After exponential phase, the growth rate decreases because the conditions changed by nutrients depletion or inhibitory by toxic waste products accumulation, and the specific growth rate becomes less than the maximum rate, $\mu_{max}$.

V. *Stationary phase*: In this phase, the rate reaches the limit of growth of the microorganism population or reaches the equilibrium state between the death and the new formation of microorganism due to nutrient exhaustion and/or inhibition. Thus, the result is no net change in numbers of microorganism population, and the specific growth rate of microorganism will be zero.

VI. *Death Phase*: In this phase, some microorganisms may remain alive and continue to metabolize, but most of them will die or lose their viability and lyse. In this case the specific growth rate of microorganism is negative, $\mu < 0$.

### 7.3.2 Temperature

Optimum condition of temperature for the functioning of cell metabolism and microorganism growth depends on the species of microorganism. Some types of microorganism can grow at temperatures of about -10 °C, and others above 100 °C. Figure 7.3 shows the different temperatures of microorganism's growth ranges [2]. In the application of biowaste treatment, microorganisms can be classified according to temperature range as following:

1. *Psycrophilic* microorganism grows at temperatures that range between -5 °C and 20 °C; the optimum temperature is about 15 °C.

2. *Mesophilic* microorganism grows at temperatures that range between 20 °C and 45 °C; the optimum temperature is about 37 °C.

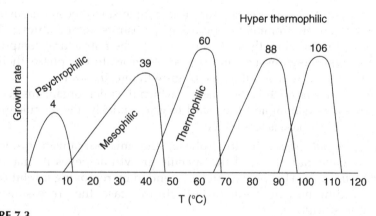

**FIGURE 7.3**

Different temperature of microorganism's growth ranges.

3. *Thermophilic* microorganism grows at temperatures between 45 °C and 70 °C with its best at about 60 °C.

### 7.3.3 pH

The variation in pH (acidity or alkalinity) influences the growth of microorganisms. Every microorganism has a pH range between 2 and 3 pH units for its growth. Generally, most natural environments have a pH between 4 and 9. Few species, however, can grow at pH values of greater than 9 or lower than 3.

Microorganisms that grow below pH 5.5 are called *acidophiles*, while those that grow at pH values in the range of 5.5 to 7.9 are called *neutrophils*. Microorganisms that grow at pH 8 or higher are called *alkaliphiles*.

### 7.3.4 Biochemical Oxygen Demand (BOD) and Chemical Oxygen Demand (COD)

Oxygen demand (OD) is an important parameter in bioprocessing because $O_2$ is necessary to aerobic microorganism during its growth and it is a limiting substrate in aerobic fermentations. The classification of microorganism according to $O_2$ consumption is as following:

1. *Aerobes* can grow at full oxygen concentration as in air (21% $O_2$) or more; they can be divided into two types:
   i. Microaerophiles that can use $O_2$ at concentration levels less than in air;
   ii. Facultative aerobes can grow under either condition (with oxygen or with very little amounts);
2. *Anaerobes* cannot grow in the presence of oxygen; they can be divided into two types:
   i. Aerotolerant anaerobes can tolerate the presence of $O_2$ but cannot use it;
   ii. Obligate anaerobes: they cannot respire oxygen (inhabited by $O_2$).

Oxygen solubility in the fermentation broth is necessary to culture conditions for aerobic microorganism; this is because the available dissolved oxygen (DO) is consumed by aerobic microorganisms that is necessary for their lives. The oxygen solubility depends on the temperature of the liquid. The concentration of dissolved oxygen in liquid decreases with the progress of time because of oxygen consumption by microorganism; this can be measured by a DO meter.

In fact, BOD is a measure of the oxygen used by microorganisms to decompose organic material present in the liquid mixture (in wastewater treatment called Mixed Liquor Suspended Solids, MLSS). The BOD can be determined by taking a sample with known quantity of dissolved $O_2$, then placing it in a sealed bottle, incubating it in the dark and determining the residual oxygen at the end of incubation. The period of incubation is usually 5 days at 20 °C; in this case, BOD is called $BOD_5$. Sometimes, the test of BOD is measured over a period of incubation more than 5 days, such as 15 or 20 days.

The microorganisms decompose the organic nitrogen compound, such as urea ($NH_2.CO.NH_2$), into ammonia ($NH_3$ and $NH_4^+$ in ionized form); this process is known as ammonification which produces nitrite ($NO_2^-$) and nitrate ($NO_3^-$) according to the following reactions:

$$\text{Ammonification reaction: } NH_2.CO.NH_2 \rightarrow H_2O + 2NH_3 + CO_2$$

$$\text{Nitrification, 1st step: } NH_4^+ + 3/2O_2 \rightarrow NO_2^- + 2H^+ + H_2O$$

$$\text{Nitrification, 2nd step: } NO_2^- + 1/2O_2 \rightarrow NO_3^-$$

The total quantity of BOD can be divided into nitrogenous and carbonaceous, in this case the BOD known with ultimate ($BOD_{ult}$). The $BOD_{ult}$ can be estimated as:

$$BOD_{ult} = a(BOD_5) + b(TKN) \tag{7.3}$$

Where: $a$ and $b$ are constants, and *TKN* is the total Kjeldahl nitrogen (organic nitrogen plus ammonia, in mg/l or mgN/L).

In general, the biodegradability of organic waste can be estimated mathematically [6] by:

$$BOD(t) = Lo(1 - e^{-kt}) \tag{7.4}$$

Where:

BOD(t) = amount of oxygen required by the microorganisms at any time t (mg/l or $mgO_2/l$),

Lo = ultimate carbonaceous oxygen demand (mg/l),

k = deoxygenation rate constant ($days^{-1}$), and

t = time (days).

The following curve (Figure 7.4) shows the $BOD_{ult}$ in function of time; the curve turns sharply upward after about 5 days; this discontinuity represents the microorganisms' demand for oxygen necessary to decompose nitrogenous materials.

The chemical oxygen demand (COD) is another parameter that provides a measure of organic material in a mixture. It uses a strong oxidizing agent,

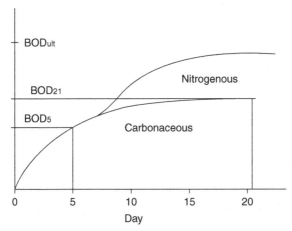

**FIGURE 7.4**

Typical long-term ultimate BOD for organic material decomposition.

such as acidic potassium dichromate ($K_2Cr_2O_7$) to oxidize the organic matter to $CO_2$.

The result of COD test is always higher than that of BOD test because of the COD test oxidation excess. Also, almost all organic compounds can be oxidized, while only some of organic compounds can be decomposed during the BOD test. For example, in case of degradation of cellulose, the result of COD is more than that of BOD. In addition, the COD test has a short time of about 2 hours for test completion or less than that in case of using modern instruments. Frequently, both tests are used to estimate the amount of bio-degradable and non-biodegradable organic matters.

### 7.3.5 Characteristic of Feedstock (Carbon/Nitrogen Content)

The characteristic of the organic materials, such as carbon/nitrogen con-tent, is an important factor for biodegradability and control of process. For example, in the case of decomposition of the organic materials by aerobic microorganism in the composting process, the initial value of C:N ratio is an important factor in the success of composting; the optimal value is about 30:1. In the case of anaerobic digestion (AD), the C:N ratio is an important factor for production of biogas, and in controlling the AD process. In this case, high C:N ratio produces a high acid content and low methane produc-tion [7], but the organic acids can be converted to methane by methanogens as motioned in the description of AD process. The C:N ratio determines the biodegradable fraction. The values of C:N ratio is variable with the type of the organic material. Table 7.2 provides examples of organic materials and their related C:N ratio [8].

**TABLE 7.2**

The C:N ratio for some organic materials (biowaste)

| Material | C:N |
|---|---|
| Food wastes | 15:1 |
| Sewage sludge | (digested) |
| | 16:1 |
| Grass clippings | 19:1 |
| Cow manure | 20:1 |
| Horse manure | 25:1 |
| Leaves and foliage | 60:1 |
| Bark | 120:1 |
| Paper | 170:1 |
| Wood and sawdust | 500:1 |

### 7.3.6 Moisture Content

Moisture content is an important factor for biological processes. The wet environment is necessary for major types of microorganism activities and for facilitating the operation of processing plant. This factor is called humidity or wetness, and it can be estimated by percentage. The calculation of moisture (water) content can be obtained by the following equation:

$$Moisture\% = \frac{\left[\left(Total\,weight\,of\,organic\,matter - dry\,weight\,of\,organic\,matter\right)\right]*100}{Total\,weight\,of\,organic\,matter}$$

(7.5)

The optimum moisture content of composting process is around 60%. In the case of AD, there are two main methods used; these are dry anaerobic digestion (24–40% total solids) and wet anaerobic digestion (10–15% total solids) [9]. It can be said that if the content of total solid (TS) is more than 15%, the process will be dry, while if the TS is less than 15% the process is wet. In case the TS content is higher than 30%, the operation will be complicated, and several operational problems will be faced such as malfunction of the pumping process, difficulties in stirring and/or stacking in feedstock. In this case, addition of external liquids (water) to the AD plant becomes inevitable, although less water is needed to accelerate the heating of the digester unit and, thus, produce more gas per feedstock unit.

### 7.3.7 Additional Factors

There are several other factors that influence the conversion calculation of biomass in biological processes; these include:

*Organic loading rate* (OLR): OLR is an important factor for determining the appropriate size of the bioreactor; this depends on several factors including kinetics of reactions, types of biomass and bioreactor; this rate can be expressed as:

$$OLR = C_iF/V \qquad (7.6)$$

Where:

$C_i$ = influent biodegradable COD concentration (mg/l),

F = flow rate (m³/day),

V = bioreactor volume (m³).

*Biomass yield coefficient (Y)*: Y is a quantitative measure of cell growth in biological processes; it is expressed as:

$$Y = \Delta X/\Delta S \qquad (7.7)$$

Where:

X = concentration of volatile suspended solids (VSS), measured in mg/l;

S = concentration of substrate (consumed of chemical oxygen demand (COD) or BOD) (mg COD/L).

The specific biological activity of biomass necessary to utilize the substrate can be expressed as:

Specific substrate utilization rate = COD removed/(VSS .day)

*Hydraulic Retention Time* $(\theta_H)$: It is defined as the duration for the biomass (biowaste) to remain in the reactor. It is used to calculate the reactor volume (volume = flow rate × $\theta_H$).

*Solids Retention Time* $(\theta_c)$: The time necessary for the stabilization of biomass (biowaste). Both parameters, $\theta_H$ and $\theta_c$, are necessary in the design of biological treatment processes.

## 7.4 Anaerobic Digestion

Anaerobic digestion (AD) is a decomposition of organic materials by anaerobic microorganisms (known as anaerobes) into solid, liquid and gas (called biogas that is rich in methane and carbon dioxide). This process passes in three stages:

1. *Hydrolysis or liquefaction stage*: Khanal (2008) presented a graphical representation of the conversion steps of complex organic materials into different groups of bacteria [10] which are involved in AD process. In this stage, the microorganisms (hydrolytic fermentative bacteria) such as *Clostridium* and *Peptococcus* can degrade carbohydrate, cellulose, lignin, pectin, starch, protein, fats and lipid compounds into soluble organic compounds such as sugars, fatty acids, amino acids or organic acids and gas as carbon dioxide, hydrogen and ammonia. For example, cellulose is degraded into glucose by bacteria, the reaction is catalyzed by the enzyme *cellulase* as in animal rumen. In this phase, the concentrations of produced gas from the fermentation can rise to about 80% carbon dioxide and 20% hydrogen [7].

2. *Acidification stage or acidogenesis and acetogenesis stage*: Some references divided this stage into two additional stages; these are the acidogenesis and the acetogenesis. The soluble organic compounds formed in the first stage is degraded by acetogen microorganisms (acetogenic bacteria) such as *Syntrophobacter* and *Syntrophomonas* into simpler compounds such as alcohols and acetone, propionic acid, acetic acid and its derivatives and carbon dioxide and hydrogen. As example, the acidogenesis can convert protein into volatile fatty acids such as propionic and butyric acid. In this phase, the concentrations of carbon dioxide and hydrogen decrease [7].

3. *Methanogenesis stage*: The primary substrates for methanogenesis are $H_2$, $CO_2$ and acetate. The organic acids (mainly acetic acid and certain other oxidized compounds or their derivatives) that are produced in the previous stages can be converted into methane and carbon dioxide by microorganisms (methanogenic bacteria or acetotrophic or aceticlastic methanogens such as *Methanosarcina* and *Methanothrix*). Hydrogenotrophic methanogens such as *Methanobacterium* and *Methanobrevibacterium* can convert $H_2$ and $CO_2$ into methane. Microorganisms (homoacetogens) can convert $H_2$ and $CO_2$ into acetate (synthesis of acetate). In this phase, the concentrations of biogas are about 60% methane and 40% carbon dioxide [7]. Figure 7.5 shows the principle steps of AD process.

The anaerobic decomposition can be simply represented by:

$$\text{Organic matter} + \text{microorganisms} => CH_4 + CO_2 + H_2 + NH_3 + H_2S + \text{biomass}$$

Furthermore, the general chemical equation of anaerobic reaction is given by:

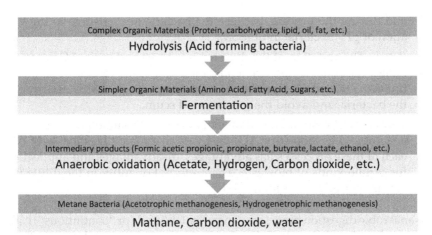

**FIGURE 7.5**

AD process: the conversion steps of complex organic materials.

$$C_aH_bO_cN_d + [(4a - b - 2c + 3d)/4]H_2O => [(4a + b - 2c - 3d)/8]CH_4 + [(4a - b + 2c + 3d)/8]CO_2 + dNH_3$$

Additional important chemical reactions of AD process are provided below [11]:

$$\text{Acetic acid: } CH_3COOH \rightarrow CH_4 + CO_2$$

$$\text{Methanol: } CH_3OH + H_2 \rightarrow CH_4 + H_2O$$

$$\text{Carbon dioxide and hydrogen: } CO_2 + 4H_2 \rightarrow CH_4 + 2H_2O$$

The biogas products, i.e. $CH_4$, $CO_2$ and $H_2O$ vapor and other minor gases, can be used as a fuel or as a chemical feedstock.

The anaerobic digestion process is a biological process that is controlled by the following factors:

*Temperature*: In general, digesters are operated at around *mesophilic* (35 °C) or *thermophilic* (55 °C) conditions. The constant temperature in the operation is essential for process efficiency.

*Retention time*: The degradation time depends on the character of biowaste, temperature of operation and the availability of bacteria. In the case of *thermophilic*, the retention time is about 14 to 21 days in the case of using a plug flow bioreactor [8].

*pH*: Ideal conditions for the methanogenic microorganisms are a pH range from 6.8 to 7.5 (Diaz et al. 1993; Wheatley 1990; IEA Bioenergy 1996). pH control is necessary during start-up, in overload conditions of process and during the process, because the bacteria are sensitive to the variation of pH (bacterial inhibition).

*Agitation:* The agitation of the digester is very important in AD because it makes the process homogeneous; in fact, it helps to mix the material, uniform concentration and obtain the temperature that is necessary to the bacteria, and avoid the formation of scum.

*Water content:* As mentioned above, there are two main types that use dry anaerobic digestion (dry AD) and wet anaerobic digestion (wet AD), each of these processes has advantages, disadvantages and applications for certain kinds of biowaste. The preferred quantity in the dilute process is typically <15% total solids.

The anaerobic digestion processes can be "batch" and/or "continuous"; it also could be "wet" and/or "dry". In addition, it can be operated by mesophilic or thermophilic and in a bioreactor as fully mixed or plug flow. Figure 7.6 shows the diagram of different classification of AD process.

Different types of reactors can be used for AD such as fluidized bed reactor (FBR), up flow anaerobic sludge blanket (UASB), anaerobic baffled reactor (ABR), anaerobic fixed film reactor (AFFR), continuously stirred tank reactor (CSTR), completely mixed contact reactor (CMCR) and multiphasic processes (MPP). The selection of the reactor type or process type depends on the required feedstock and technology.

In the following chapters we will discuss the applications of biological processes, such as composting in case of aerobic process, landfill and anaerobic digestion in case of anaerobic process, also, the technologies of physical–biological treatment and the thermal processes.

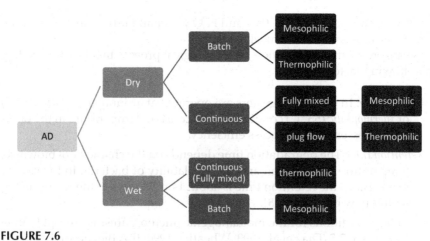

**FIGURE 7.6**

Different classifications of AD process.

## Review Questions

Define the following: aerobic process, anaerobic process, BOD, COD.
Give 4 examples on physical and chemical processes.
Discuss the main factors that control biological processes.

## References

1. V. Ivanov, Microbiology of Environmental Engineering Systems, in L.K. Wang et al., (eds.) Handbook of Environmental Engineering, Volume 10: Environmental Biotechnology: DOI: 10.1007/978-1-60327-140-0_2 © Springer Science + BusinessMedia, LLC 2010.
2. M.T. Madigan, J.M. Martinko, D.A. Stahl and D.P. Clark, 2012, "Brock Biology of Microorganisms", Thirteenth Edition, Benjamin Cummings, ISBN-13: 978-0-321-64963-8.
3. W. Reineke, 2001, "Aerobic and Anaerobic Biodegradation Potentials of Microorganisms", Chapter 1 in B. Beek (ed.), "The Handbook of Environmental Chemistry", Vol. 2 Part K "Biodegradation and Persistence" Springer-Verlag, Berlin, Heidelberg.
4. M.R. Tredici, 1999, "Bioreactors, Photo" in M.C. Flickinger and S.W. Drew (eds.), "Encyclopedia of Bioprocess Technology: Fermentation, Biocatalysis, and Bioseparation", Volumes 1 5, Wiley, ISBN 0-471-13822-3.
5. S. Katoh and F. Yoshida, 2009, "Biochemical Engineering", Wiley-VCH Verlag GmbH & Co. KGaA, Weinheim, ISBN: 978-3-527-32536-8.
6. Madigan, Martinko, Stahl and Clark, "Brock Biology of Microorganisms".
7. R.E. Weiner and R.A. Matthews, 2003, "Environmental Engineering", 4th Edition, Butterworth-Heinemann, Elsevier Science (USA), ISBN: 0750672943.
8. P.T. Williams, 2005, "Waste Treatment and Disposal", Second Edition, Wiley, ISBN 0-470-84912-6.
9. G.M. Evans and J.C. Furlong, 2003, "Environmental Biotechnology: Theory and Application" Wiley, ISBN: 0-470-84372-1.
10. L. Luning, E.H. van Zundert and A.J. Brinkmann, 2003, "Comparison of Dry and Wet Digestion for Solid Waste", Water Science & Technology, Vol. 48, Supplement 4, pp. 15–20.
11. S.K.Khanal,2008,"AnaerobicBiotechnologyforBioenergyProduction:Principles and Applications", Wiley-Blackwell. ISBN-13: 978-0-8138-2346-1.
12. Williams, "Waste Treatment and Disposal".
13. Williams, "Waste Treatment and Disposal".
14. Williams, "Waste Treatment and Disposal".
15. J.R. Taricska, D.A. Long, J.P. Chen, Y.-T. Hung and S.-W. Zou, 2007, "Anaerobic Digestion" in L.K. Wang, N.K. Shammas and Y.-T. Hung (eds.), "Biosolids Treatment Processes, Handbook of Environmental Engineering", Vol. 6, Humana Press Inc. Totowa, New Jersey, ISBN 978-1-58829-396-1.
16. Evans and Furlong, "Environmental Biotechnology: Theory and Application".

# 8

## Biowaste Solid Materials Treatment

*Key Learning Objectives*

- Understanding the application processes of biowaste.
- Understanding biowaste disposal.
- Understanding the terms composting, landfilling, anaerobic digestion and the technologies of physical–biological treatment.

## 8.1 Introduction

In this chapter, we will discuss the main application processes of biowaste treatment and disposal. Engineering technologies include several processes for biowaste treatment such as landfilling, composting, anaerobic digestion and BTA processes. These processes are biological processes (see Chapter 7).

## 8.2 Landfilling

Biowaste materials discarded in landfills gradually undergo a natural process of biodegradation (anaerobic process) with speed depending on several factors that control the biological process such as nature of waste, temperature and moisture content.

The biowaste materials that are delivered to a landfill start to decompose; this will increase the levels of leachate and methane gas produced. Initially, it begins to break down into smaller components and carbon dioxide ($CO_2$) and other gases. The degradation process takes place firstly in an aerobic environment in which the content of $O_2$ is embroiled between waste materials and surface. The next phase takes place in an anaerobic environment since the content of $O_2$ has been consumed. The mechanism of degradation process is very complex and passes through several intermediary reactions. Leachate and several gases are produced by this process:

$$\text{Organic material} \rightarrow CH_4 + CO_2 + H_2 + NH_3 + H_2S + H_2O.$$

The mean types of landfill gases (LFG) are methane ($CH_4$) of about 45–60% and carbon dioxide ($CO_2$) of about 35–45% [1]. However, several additional trace gases of varying chemical composition are raised, such as hydrogen, nitrogen, oxygen, water vapor and other trace constituents.

The leachate is a water that is rich with pollutants with high concentrations of heavy metals and other contaminants. The percolating leachate passes through the waste materials and tends to leach out both organic and inorganic substances; this could cause the contamination of the groundwater. The pollution severity of landfill leachate is very complex since it is controlled by several factors; these are the origin of leachate contents, age of waste materials and management procedures applied at the landfill. The effect of leachate reaches its maximum during the first two years of the landfill age.

Legislations in many countries enforce the collection of gases produced in landfills and their use for energy generation, if possible, or be flared. These procedures aim to prevent or reduce the negative environmental effects of landfills, especially hazardous waste materials. The legislation related to pretreatment of soil employs using leachate collection equipment system, pumping and treating it. In case of groundwater protection, liner specifications as an impermeable clay or geopolymer barrier are used. In addition, extensive studies to investigate potential health risks within 2–3 kilometers radius of the landfills must be made.

Recently, sanitary landfills are introduced. In this type of landfills, waste materials are deposited in a systematic mode and built into layers of limited space; this is the smallest volume which is covered with a layer of earth at the end of waste delivered and called "cell" corresponding to a day's waste arrived. For more information about the landfill construction techniques, see Baird and Cann (2012) [2] and Everett (1999) [3].

Landfills can also be classified according to topography of location or the technology employed in the operation of the landfill into *open and closed pit landfill*. The open pit landfill is positioned in a narrow gorge with steep sides having one side open, and drainage of leachate occurs by gravity because the bottom of the pit is sloped.

The closed pit landfill has a base surrounded by terrain and the leachate must be constantly pumped out from the bottom of the pit. The gases can easily escape out from uncontrolled covers; this depends on the porosity of the cover.

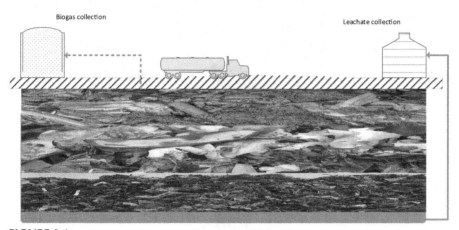

**FIGURE 8.1**

Illustration of a typical pit landfill of an anaerobic bioreactor.

Furthermore, landfills can be constructed on hills/slopes or in a valley. In this case, it is necessary to conduct extensive studies of the topographical and hydrogeological conditions to facilitate the operation and monitoring of the site. This is also imposed by legislation in order to minimize the risks to the environment and human heath when selecting the landfill site.

A pit landfill consists of two units that are used for landfill gas treatment and leachate treatment. It is an anaerobic bioreactor as illustrated in Figure 8.1.

The land disposal of hazardous waste must be systematic in secured landfill as in Figure 8.2.

### 8.2.1 Pollution Indicators of Leachate Composition

In addition to microorganism population, the main indicators of organic pollution are COD, $BOD_5$, TOC, main inorganic ions, such as Cl and $SO_4$, and heavy metals. A typical BOD of a landfill leachate is about 20,000 mg/L [4] while it is about 10,000 mg/L for the $BOD_5$ (Table 8.1).

### 8.2.2 Leachate Calculation

The percent of moisture can be calculated in laboratory using the following formula:

$$\text{Percent Moisture} = 100 - (\text{dry weight/wet weight} * 100) \quad (8.1)$$

Practically, however, material balance method is applied in which the total quantity of water in the landfill is calculated using the quantity of water entering the topsoil layer and the quantity that gets out from the landfill pit.

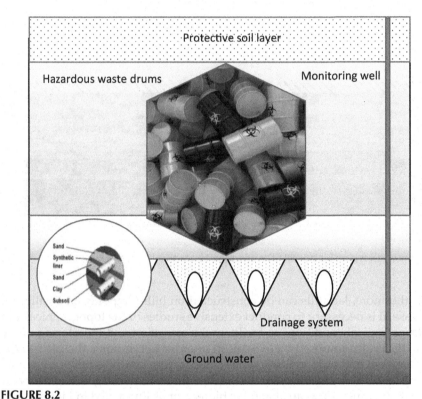

**FIGURE 8.2**

An illustration of a secured landfill used for hazardous waste disposal.


The water balance of a landfill is described by the balance equation (8.2) of climatic leachate formation [5]:

$$L_F = P - V_E - V_T - E_S \qquad (8.2)$$

Where:

$L_F$ = climatic leachate formation,

P = precipitation, controlled water addition, if required,

$V_E$ = evaporation,

$V_T$ = transpiration,

$E_S$ = effluent surface water.

Leachate effluent at the landfill base is given by:

$$E_B = L_F - S \pm R \pm W_D + W_C \qquad (8.3)$$

**TABLE 8.1**

Typical values for leachate composition[a]

| Constituent (All units are in milligrams per liter except pH) | Range | Typical |
|---|---|---|
| BOD$_5$ (5-day biochemical oxygen demand) | 2,000–30,000 | 10,000 |
| TOC (total organic carbon) | 1,500–20,000 | 6,000 |
| COD (chemical oxygen demand) | 3,000–45,000 | 18,000 |
| Total suspended solids | 200–1,000 | 500 |
| Organic nitrogen | 10–600 | 200 |
| Ammonia nitrogen | 10–800 | 200 |
| Nitrate | 5–40 | 25 |
| Total phosphorus | 1–70 | 30 |
| Ortho phosphorus | 1–50 | 20 |
| Alkalinity as CaCO$_3$ | 1,000–10,000 | 3,000 |
| pH | 5.3–8.5 | 6 |
| Total hardness as CaCO$_3$ | 300–10,000 | 3,500 |
| Calcium | 200–3,000 | 1,000 |
| Magnesium | 50–1,500 | 250 |
| Potassium | 200–2,000 | 300 |
| Sodium | 200–2,000 | 500 |
| Chloride | 100–3,000 | 500 |
| Sulfate | 100–1,500 | 300 |
| Total iron | 50–600 | 60 |

[a] Glysson, et al., "Solid Waste".

Where:

$E_B$ = leachate effluent at the landfill base (into a drainage system, or underground if no bottom sealing exists),

S = storage,

R = retention,

$W_D$ = water demand or release by biological conversion,

$W_C$ = consolidation.

### 8.2.3 Landfill Hydrology Model

Landfill hydrology model is used to estimate the hydrologic characteristic conditions of the landfill; it is used also to design landfill bioreactors. The computer program Hydrologic Evaluation of Landfill Performance (HELP) [6] is a typical software that can be used to estimate the water balance under different design scenarios [7]. This model includes three types of input data: (a) climatological data (evapotranspiration, precipitation data, temperature and solar radiation data); (b) soil data (soil–material interfaces and properties for hydraulic conductivity, wilting point, field capacity and porosity) and (c) design data (landfill liner system cross-sections including vertical percolation layer, lateral drainage layer, soil layer and geomembrane liner) [8]. Darcy's Law is used for calculating leachate flow rate that travels through the waste mass:

$$Q = kiA \tag{8.4}$$

Where:

Q = flow rate ($l^3$/t) into landfill,

k = permeability of media (l/t),

i = hydraulic gradient (unitless),

A = cross-sectional area ($l^2$).

Landfills design depends on several factors such as political and social constraints, legislation, study of area and site selection (as seismic impact zones, topography, soil conservation and land-use maps for effects on surrounding environment) and quantity of waste and lowest cost per ton of waste disposal with trucking. Landfill design should include the waste in safe manner with established minimum site size, for example a minimum depth of solid waste of 6 m and a minimum life of 10 years. Further readings about landfills can be found in the collection of knowledge around sustainable sanitation and water management on the web with open source copyright in the site of Sustainable Sanitation, Water Management & Agriculture (SSWM): www.sswm.info/content/landfills.

## 8.3 Composting

Composting is a biological stabilization of organic waste by aerobic conditions. It is a natural way of waste treatment to convert waste into soil without environmental effects. Composting can be applied to semisolid and solid organic wastes from farms and households such as animal manures, agricultural residues, human excrement (night soil), sludge and municipal refuse (usually higher than 5% organic wastes [9]). The final product of composing is an organic material called "humus" or "fertilizer" used as land reclamation (plant mulch or soil conditioner) in horticultural or agricultural applications. The final product of composting makes the soil cultivation an easy task and reaches the plants for nutrition purposes.

Composting is an aerobic decomposition of organic wastes in the presence of oxygen with the help of microorganism (mainly bacteria, actinomycetes and fungi). The biological metabolism of microorganism can be produced as final-product water, carbon dioxide ($CO_2$), $NH_3$ and heat [10]. This is an

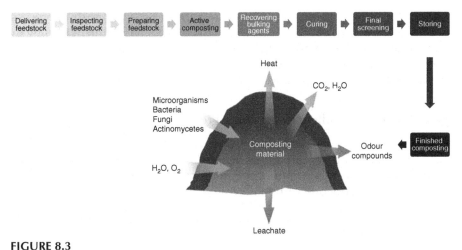

**FIGURE 8.3**

Process of composting and its steps at industrial levels (the final figure is adapted from Public Works and Government Services of Canada).

exothermic reaction that produces heat which increases with the addition of oxygen (in aerated condition). At industrial level, the composting passes through several steps as illustrated in Figure 8.3.

Next, the delivered feedstock passes to inspection phase to be controlled for the absence of non-degradable items as plastic materials, glass materials and metal cans. In the preparing step, the feedstock passes into physical treatment such as grinding or shredding and screening then mixed with other materials and water to make it homogeneous and ready for the next phase. In the active composting phase or step, the microorganisms are activated to decompose the organic materials; this phase takes different intervals, which depend on the utilized technology, and methods of composting applied (known as time of composting from few weeks to many months).

Sometimes, the compost is covered with bulking agents such as woodchips that enrich the feedstock with carbon that is considered as nutrients to the microorganisms during the biological activity or curing phase, but this increases the anaerobic conditions. At the end of this phase, the compost becomes fully stabilized and matured. Finished compost can be screened, sorted (final screening and storing) and dried; in this case, the product is considered as a finished product (industrial product) which will be ready to be sold or stored.

The stabilization of organic waste to be completed depends on the processing method; it requires several days in case of mechanical methods or several weeks in case of spreading methods. In certain cases, 3 to 6 months or more are required, depending upon the climatic conditions of the zone and the utilized method or operation process. In the case of natural aeration without turn, the decomposition has slow rate and the composting process is called slow or cold composting. In case the process has frequent turning to provide aeration, the composting is called fast or hot composting. Sometime, the composting is an anaerobic decomposition when organic waste is covered in a hole.

## 8.3.1 Factors Controlling Composting Process

Composting process is controlled by several factors; these are: temperature, moisture content, carbon/nitrogen (C/N) ratio and the amount of air (oxygen) to make an optimum environmental condition for the waste decomposers. A brief description is provided below.

### 8.3.1.1 Temperature

The activity of microbial population changes during composting because the biological process produces heat then changes the temperature combining both mesophilic and thermophilic activities. At ambient temperature, the activity of microbial starts with latent phase, then it enters growth and reproduction phases, known as mesophilic stage; the temperature is between 25 and 45 °C. At the end of these phases, the temperature reaches its maximum (does not exceed 70 °C); this is known as thermophilic phase (see Chapter 7).

High activity of microorganisms during the thermophilic phase, which increases the composting process, and the heat produced during this phase destroy pathogenic organisms and stabilized organic material (known as pathogen sterilization). The optimum temperature for the thermophilic composting is in the range of 50–60 °C. The temperature is gradually dropped to the mesophilic phase; this is known as maturation phase or secondary mesophilic phase. The temperature continues to decrease until it reaches the ambient temperature [11]. Figure 8.4 shows the effect of temperature, heat lost and relationship between microbial growth and temperature during the composting phases.

### 8.3.1.2 Moisture Content

Water content (or moisture) is an important factor for biological decomposition of the organic waste. The water content could reach 100% for composting [12], but the optimum range is around 60% [9]. The suitable range of moisture content is between 50 and 70%. If it decreases to less than 20%, the biological process will be inhibited, and consequently, causing reduction of microbial activity. Due to the high temperature inside the piles (thermophilic period) and surface evaporation, some water needs to be added during the continuous composting time to maintain an optimum environment for control of composting operation.

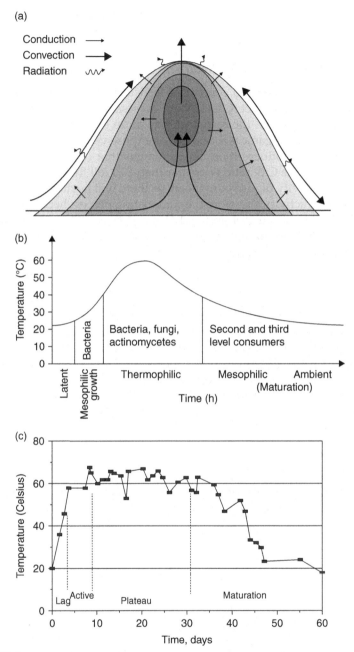

**FIGURE 8.4**

Illustration of the effect of heat during composting time. (A) Mechanisms of heat loss from a thermophilic compost pile, adapted from Trautmann and Krasny, (1997) (N. Trautmann and M. Krasny, 1997, "Composting in the Classroom: Scientific Inquiry for High School Students", eBook available online at http://cwmi.css.cornell.edu/compostingintheclassroom.pdf on 2/9/2015); (B) Phases of composting process in compost piles relationship between microbial growth and temperature, adapted from Chongrak Polprasert (2007) (see Polprasert, "Organic Waste Recycling: Technology and Management"); (C) Typical temperature curve during composting time, adapted from Luis Diaz et al. (2002) (see L. Diaz et al., "Composting of Municipal Solid Wastes").

### 8.3.1.3 Carbon/Nitrogen (C/N) Ratio

The carbon to nitrogen ratio content of feedstock is an important factor in biological treatment. In fact, C and N are necessary for the microorganism's metabolic activity. When the feedstock has more N content or C:N ratio is lower than 20:1, the organic waste will decompose very rapidly, and the process of treatment produces more ammonia gas that causes odor problems. Vice versa, when the C:N ratio is greater than 20:1, all the N will be used for the metabolic activity of microorganism, or there will be no enough nitrogen to the metabolic activity, therefore the composting process becomes slow. The range of C:N is between 25:1 and 35:1 [13] and the optimum value is about 30 to 1. When the carbon to nitrogen ratio is around the optimum value, the composting process will run more effectively [14].

### 8.3.1.4 Particle Size

Particle size of the materials plays an important role in determining the surface area of the materials available to microbial attack or availability of carbon and nitrogen. The smallest particle size feedstock has the advantages of speeding up the decomposition process of composting and preventing aeration from reaching all parts of the pile. Large-particle size, on the other hand, provides the compost pile with enough porosity. The optimum particle size of feedstock for composting is variable between 1 and 5 cm [15]. Figure 8.5 shows the effect of particle size on composting.

### 8.3.1.5 Aeration

Aeration is an essential factor for composting process (see OD or BOD and COD factors in Chapter 7). It is an air introduced into material mass in direct

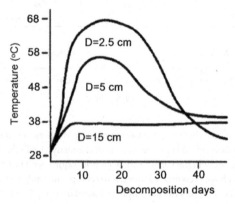

**FIGURE 8.5**

Typical carves showing the effect of particle size on composting process.

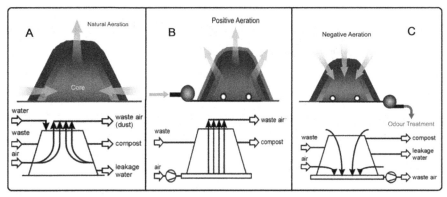

**FIGURE 8.6**

Methods of aeration system: (A) natural aeration, (B) forced aeration (positive) and (C) forced aeration (negative). Adapted from Frank Schuchardt (2005) (Schuchardt "Composting of Organic Waste") and Public Works and Government Services of Canada (2013) (Public Works and Government Services of Canada (PWGSC), 2013, "Technical Document on Municipal Solid Waste Organics Processing" Cat. No.: En14-83/2013E, ISBN: 978-1-100-21707-9).

or indirect modes. Stoichiometrically, the average of OD for batch process is 2 kg $O_2$/kg of volatile solids or a flow rate between 12 and 30 $m^3$ air/hour per kg of dry mixture. In the continuous processes, it reaches to 43 kg of air per kg of mixture or a flow rate of 1200 $m^3$ air/hour per dry ton [16]. Aeration system depends on the methods of composting as illustrated in Figure 8.6. There are natural and forced aeration; the forced aeration can be positive or negative.

## 8.3.2 Methods of Composting

Over the 20th century, the composting process has experienced significant technological developments worldwide, in particular, in China, India, Europe and the US [17]. Currently, there are several technologies that are used in composting organic wastes at industrial levels; these are mainly classified in three categories [18]: (a) windrow, (b) static pile and (c) in-vessel. Furthermore, these can be classified in two general systems, namely the open system and the closed/reactor or "in-vessel" system (Figure 8.7). Both open and enclosed systems can be conducted at two scales: home or industrial. At home, it is known as *home composting*. The selection of the application of each method depends on the quantity of feedstock, capital investment and time necessary to end process.

*Home composting*: It is also called method of farming. The aims of composting in home are stabilization, drying and sanitation of organic waste, mass and volume reduction of waste, elimination of bad odor and creating rich humus for garden or lawn.

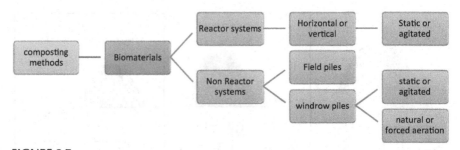

**FIGURE 8.7**

Illustration diagram showing composting methods.

It is a traditional system for dryer waste used by farmers in villages. In this system, waste materials (animal manure or garden waste) are collected in the form of a growing pile. It has a natural aeration and needs sometimes before stabilization and be used as a land fertilizer. It could take one year or more to be ready to be used. The maximum pile volume that ensures enough natural aeration is about $1m^3$. Sometimes, the waste materials are collected in-vessel (in bin) either as open bins or enclosed bins. The following figure shows the different types of home and industrial composting techniques (Figure 8.8).

Composting in *enclosed bins* uses an enclosed container such as a barrel or tumbler. It is a small scale of the industrial technique that is used at homes, and it is considered a practical and easy composting method. However, composting could take some time to produce compost that ranges from several weeks to several months. The method needs, occasionally, turn, such as running top to bottom or laid on side and rolled or turned at the axe of drum/tumbler, in-vessel (Figure 8.8A).

*Anaerobic composting* at homes is conducted by digging a trench in the earth or a hole which will be used for burying waste materials (including dead animals). It is also known as *"trench composting"* or *"pit composting"*. This technique has the advantage of reducing odors. However, longer time is required to compost compared to the aerated composting technique; this might take up to 6 months or more. This method is used for degradation of animal carcasses, sometimes worms are used to speed the process when the composting is done in reactor or vessel. This method is known as *Vermicomposting* (vermi is Latin word that means worm. See Figure 8.9).

*Industrial composting:* The industrial operations have added high value to the biowaste materials and, consequently, its expense. However, since the production is made in large quantities, it is usually commercially convenient. The large-scale composting or industrial levels are included both in open and enclosed systems such as passive piles, windrows, aerated static piles and in-reactor systems (Figure 8.8B).

Open systems are conducted in piles or windrows. Piles or heap composting is a simple static pile; the organic mass is collected in a pile. Usually, it has

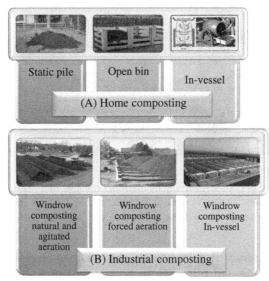

**FIGURE 8.8**

Examples of product composting (A) at home and (B) industrial.

**FIGURE 8.9**

Photo of a worm used in vermicomposting (adapted from the free web) (pikist.com, free domain, available online at www.pikist.com/free-photo-vqbsh, photo: https://p0.pikist.com/photos/488/563/worm-vermiculture-humus-earth-soil-rico-nutrients-compost-composting.jpg, on 20/06/2020).

natural ventilation or aeration, in this case, the composting takes a long time which could reach one year or more to produce the final product. If the aeration is forced in, the biomaterial degrades in less time. The forced aeration methods are positive when the air enters the pile (positive aeration) or negative aeration when the odor comes out (passive aeration), but in this case the flow gas must be treated (see Chapter 11). Usually, the system of forced aeration uses fans as in Figure 8.6.

**FIGURE 8.10**

Windrow composting periodically turning by machines (images adapted from the free web) (pxfuel.com, free domain, available online at www.pxfuel.com/en/free-photo-epapy, and www.pxfuel.com/en/free-photo-epzkl, on 20/06/2020).

*Windrow composting*: It refers to the biowaste materials that are piled up in rows or elongated piles. It can be static or movable or periodically turned by machines or screened; in this case it is called active composting (Figure 8.10).

Usually, a strong odor exists, in particular, when the raw materials are mixed with sludge. Windrow composting needs less time to produce the final compost. Due to the continuing aeration and turning which helps in speeding the biological reactions, this method takes about one month to produce the compost. Sometimes, in the case of *static composting*, or mechanical aeration, the airflow enters through perforated pipes under the composting materials.

In the case of natural aeration, the process is called *passive aeration*. This is caused by the transfer of heat by convection from the center to the top of the pile which creates a vacuum in the center of the pile; this vacuum causes fresh air to enter from the sides of the pile.

On other hand, the industrial composting in closed systems is faster, need more operations and maintenance and the final product is more costly compared to home composting. These systems can be static or removable; also, the position of reactors can be vertical or horizontal. For more information about industrial composting and equipment used in industrial operations, see Rynk (1992) [19].

## 8.4 Physical–Biological Processes

Mechanical biological treatment (MBT) or biological mechanical treatment (BMT) is a process or a system that combines physical and biological treatment of waste processing facilities. This system can be enabled to recover materials

**FIGURE 8.11**

Scheme of biowaste treatment by BTA processes.

contained within the mixed waste and facilitate the stabilization of the bio-degradable component of the material, such as the processes of composting or anaerobic digestion.

Different types of waste can be used with this system, such as municipal solid waste, commercial and industrial waste and sewage sludge. These are used for producing renewable fuel (biogas), digesting organic fertilizers for soil improvement purposes and recovering recyclable materials such as metals, paper, plastics and glass. The following diagram (Figure 8.11) illustrates the biowaste treatment by anaerobic digestion (AD) process.

An example of the MBT system is the BTA processes; this is described in the following paragraphs.

*BTA Process* is a wet pretreatment technique that was developed in 1980 by two German waste management companies, BioTechnische Abfallverwertung GmbH (BTA) [20] and Müll und Abfalltechnik GmbH (MAT) [21]. In 1993, the two companies started the operation of a plant with a capacity of 5,000 t/y of biowaste treated in the city of Baden-Baden, Baden-Württemberg, Germany [21]. The process is based on two major steps: (a) mechanical pretreatment and (b) biological conversion. It is a closed loop technology that integrates sophisticated wet pretreatment techniques with proven methods of anaerobic digestion (AD). It can transform biowaste materials such as organic fraction of municipal solid waste from households, commercial and agricultural waste into high-grade biogas and valuable compost. The biogas produced from anaerobic digestion can be converted into electrical energy for using a combined heat and power system by engines.

In the phase of pretreatment, the BTA process separates the organic matters with high efficiency from the rest of the waste materials (non-biodegradable, such as plastics, textiles, stones, metals) using a hydro-mechanical technology.

The products of BTA process are (a) recyclable materials, such as plastics and metals; (b) organic material destined into composting or anaerobic digestion; (c) materials that have higher caloric value destined into combustion process for production of fuel and (d) residual wastes or rejects destined into landfill.

In general, BMT has several advantages; these include reducing the quantity of residual waste materials and increasing its compatibility, decreasing the quantity of waste materials that go to landfills thus reducing the leachate concentrations and gas rises from landfill. In addition, it reduces the active potential of native organic substances. During these phases, the received MSW are subjected to a dry pretreatment followed by sorting and rejection of overflow. The sorting process separates the waste materials into recyclable materials and combustibles known as refuse derived fuel.

Biowaste enters a screw mill where pretreatment occurs. Next, the biowaste materials enter into hydropulper and grit removal system where non-biodegradable fractions, such as light fraction as plastics and heavy fraction, as metals, bones, batteries stones and glass fragments, are removed.

After the pretreatment, the biowaste materials become cleaned, homo-genous pulp and ready to be sent to digestion (bioreactor). After the phase of anaerobic fermentation, the produced biogas is sent into a combined heat and power system, while the rest of solid materials are sent into mechanically dewatered then sent into composting. The liquid fraction is partially recycled in the plant and the other part is used as liquid fertilizer.

The multistage fermentation is better than one stage fermentation pro-cess, because it reduces the time of the process about 50%. It has a higher biogas yield for same quantity of waste operated, but heating value of biogas produced for both systems is the same [22].

## Review Questions

How can biowaste solid materials be treated?
Define landfilling.
Describe the disposal of hazardous waste by systematic method in a secured landfill.
Describe landfill hydrology model.
What is composting?
Mention five factors for controlling composting process.
What are the differences between home and industrial composting?
What is vermicomposting?
Describe the mechanical biological treatment.

## References

1. T.E. McKone, S.L. Bartelt-Hunt, M.S. Olson and F.D. Tillman, 2011, "Mass Transfer within Surface Soils" in Louis J. Thibodeaux, Donald Mackay (eds.) "Handbook of Chemical Mass Transport in the Environment", CRC Press, Taylor & Francis Group, LLC. ISBN: 978-1-4200-4755-4.

2. C. Baird and M. Cann, 2012, "Environmental chemistry", Fifth Edition, W.H. Freeman and Company, New York, ISBN-13: 978-1-4292-7704-4.
3. J.W. Everett, 1999, "Sanitary landfills" in Irene Liu (ed.) "The Environmental Engineers' Handbook", CRC Press LLC, International Standard Book Number 0-8493-2157-3.
4. G.M. Evans and J.C. Furlong, 2003, "Environmental Biotechnology: Theory and Application" Wiley, ISBN 0-470-84372-1.
5. K. Hupe, K.-U. Heyer and R. Stegmann, 2003, "Water infiltration for enhanced in situ stabilization". Proceedings Sardinia 2003, Ninth International Waste Management and Landfill Symposium, S. Margherita di Pula, Cagliari, Italy; 6–10 October 2003, by CISA, Environmental Sanitary Engineering Centre, Italy.
6. P.R. Schroeder, T.S. McEnroe, J.W. Sjostrom and R.L. Peyton, 1994, "The Hydrologic Evaluation of Landfill Performance (Help) Model". U.S. Army Corps of Engineers. Waterways Experiment Station: Vicksburg, MS 39180-6199, Interagency Agreement No. DW21931425.
7. ITRC (Interstate Technology & Regulatory Council), 2003, "Technical and Regulatory Guidance for Design, Installation and Monitoring of Alternative Final Landfill Covers", ALT-2, Washington, D.C.: Interstate Technology & Regulatory Council, Alternative Landfill Technologies Team. www.itrcweb. org.
8. ITRC (Interstate Technology & Regulatory Council), 2005, "Characterization, Design, Construction, and Monitoring of Bioreactor Landfills", ALT-3. Washington, D.C.: Interstate Technology & Regulatory Council, Alternative Landfill Technologies Team. www.itrcweb.org.
9. C. Polprasert, 2007, "Organic Waste Recycling: Technology and Management", 3rd Edition, IWA Publishing, London, UK, ISBN13: 9781843391210.
10. PWGSC, Public Works and Government Services of Canada, 2013, "Technical Document on Municipal Solid Waste Organics Processing" Cat. No.: En14-83/2013E, ISBN: 978-1-100-21707-9.
11. L.F. Diaz, G.M. Savage and C.G. Golueke, 2002, "Composting of Municipal Solid Wastes" in George Tchobanoglous and Frank Kreith (eds.) "Handbook of Solid Waste Management", 2nd Edition, McGraw-Hill, New York.
12. F. Schuchardt, 2005, "Composting of Organic Waste" in H.-J. Jördening and J. Winter (eds.) "Environmental Biotechnology. Concepts and Applications", Wiley-VCH Verlag GmbH & Co. KGaA, Weinheim, ISBN: 3-527-30585-8.
13. See Polprasert, "Organic Waste Recycling: Technology and Management".
14. L. Cooperband, 2002, "The Art and Science of Composting: A Resource for Farmers and Compost Producers", Center for Integrated Agricultural Systems, University of Wisconsin-Madison.
15. R.D. Raabe "The Rapid Composting Method", Vegetable Research and Information Center, University of California, Division of Agriculture and Natural Resources.
16. B. Chisholm, WSU Whatcom County Extension Community Horticulture programs, (http://whatcom.wsu.edu/ch/compost.html), Composting & Recycling, Composting Fact Sheets "Compost fundamentals", Washington State University, available online at http://whatcom.wsu.edu/ag/compost/fundamentals/needs_particle_size.htm on 7/6/2015.

17. C.V. Andreoli, M. Von Sperling and F. Fernandes (editors), "Biological Wastewater Treatment Series", Vol. 6, "Sludge Treatment and Disposal".
18. L.F. Diaz, M. de Bertoldi, W. Bidlingmaier and E. Stentiford (eds.), 2007, "Compost Science and Technology", Waste Management Series; Vol. 8, Elsevier, ISBN-13: 9780080439600.
19. E.A. Glysson, et al., 1999, "Solid Waste" in R.A. Corbitt (ed.) "Standard Handbook of Environmental Engineering", 2nd edition, McGraw-Hill, ISBN: 9780070131606.
20. R. Rynk (ed.), 1992, "On-Farm Composting Handbook", NRAES-54, Cooperative Extension, Northeast Regional Agricultural Engineering Service, 152 Riley-Robb Hall.
21. R. Haines, BTA Benefits Biowaste, waste-management-world.com, available online at www.waste-management-world.com/articles/print/volume-10/issue-3/features/bta-benefits-biowaste.html on 2/4/2015.
22. B. Bilitewski, G. Härdtle and K. Marek, "Waste Management", Translated and Edited by A. Weissbach and H. Boeddicker, Springer-Verlag Berlin Heidelberg 1997, ISBN 978-3-662-03382-1 (eBook), DOI 10.1007/978-3-662-03 3 82-1
23. Bilitewski, Härdtle and Marek, "Waste Management".
24. WtERT, The Waste-to-Energy Research and Technology Council (WtERT), 2015, "Anaerobic Digestion Systems", Text created by M.Sc.-Ing. Emmanuel Serna (agriKomp GmbH), (2013-08-10), last modified by Martin Ernst (BTA International GmbH), (2014-10-15), available online in 29 August 2015 at www.wtert.eu/default.asp?Menue=13&ShowDok=17

# 9

## Waste Disposal by Thermal Processes

---

*Key Learning Objective*

- Understanding thermal disposal processes.

---

### 9.1 Introduction

Thermal processes, sometime called heat treatment processes, use the temperature or heat to break down the waste material into submaterials such as gases and ashes. The main thermal processes are combustion or incineration, pyrolysis and gasification. These are called "processes to production energy or recover energy". In fact, power production, such as heat, steam and electricity, uses these processes using various items of equipment, such as stoves, furnaces, boilers, steam turbines, turbo-generators, etc.

As an example for thermal processes, the waste of sugar factories that is known as bagasse is often used as the main source of fuel in sugar factories for sugar mills [1]. It is used in the boilers for steam production that is used to power the process. It is also used as a primary fuel source when burned in sufficient quantity; it produces sufficient heat energy to supply all the needs of a typical sugar mill with energy to spare [2].

The diagram in Figure 9.1 shows the main differences among thermal processes: incineration, gasification and pyrolysis. These differences are concentrated in the quantity of oxygen entrance to the reactor and the types of the output products. Pyrolysis takes place in the absence of oxygen and produces gas, oil and char. Gasification is a limited supply of oxygen with a quantity less than the stoichiometric reactions and it produces gas, ash and tar. Incineration or combustion, on the other hand, is a complete oxidation process with excess supply of oxygen and produces flue gas and ash (see Box 9.1).

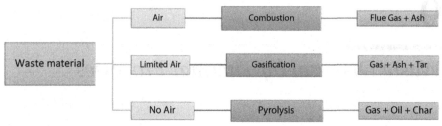

**FIGURE 9.1**

The main differences among the thermal processes: incineration, gasification and pyrolysis.

---

**BOX 9.1   BASIC DEFINITIONS**

*Combustion*: Combustion is a process of burning waste material in an incinerator where a chemical reaction occurs between a fuel and an oxidizing agent (full oxidation) that produces energy, usually in the form of heat and light.

*Burning*: Burning is a process in which a fire or a form of heat treatment is used to damage something. It results in the release of light, heat and products of combustion.

*Incineration*: The process of burning waste material completely to ashes by combustion under controlled conditions (high temperatures and oxygen levels) to reduce its weight and volume or its toxicity and often to produce energy. It is the most common thermal treatment process.

*Pyrolysis*: It is a thermochemical decomposition of organic material at a temperature range between 250 and 900 °C without oxygen.

*Gasification*: It is the conversion of organic-based carbonaceous materials into a gaseous fuel. This process takes place by reacting the material at temperatures more than 500 °C with a controlled amount of oxygen and/or steam (partial oxidation). Gasification takes place in a phase between the pyrolysis and the incineration.

---

## 9.2 Incineration

Incineration is a thermal treatment process that is used to destroy a waste material in controllable conditions of oxygen (air) and change it into submaterials that are less hazardous and safer. It is a chemical reaction (Box 9.2) and a physical process that run in the presence of oxygen in order

## BOX 9.2 CALCULATION OF COMBUSTION HEAT

A good fuel should have high calorific value when burning, chemically; a combustion reaction occurs when all substances in the waste are reacted with oxygen that then produces carbon dioxide and water. The following are examples of combustion reactions:

$$C_xH_y + [x + y/4]O_2 \rightarrow xCO_2 + [y/2] H_2O + Heat$$

$$C_xH_yO_z + [x + y/4 - z/2]O_2 \rightarrow xCO_2 + [y/2] H_2O + Heat$$

$$S + O_2 \rightarrow SO_2 + Heat$$

Where: C, $H_2$, $O_2$ and S (sulfur) are the percentage fractions by weight of each chemical constituent of the waste. In this case, it needs the ultimate analysis of waste (weight percent): %C, %H, %O, %N and %S.

For the calculation, assuming only sulfur dioxide ($SO_2$), a mixture of sulfur dioxide and sulfur trioxide ($SO_3$) are produced

## ENERGY CONTENT

The reactions are exothermic reactions, which means heat is produced that is easily distinguished when the waste is burned by flame (light and heat). The heat of combustion ($\Delta H$) is the energy released when a complete combustion takes place under standard conditions (STP).

(Standard Temperature, 273 K or 0 °C and Pressure, 1 atm or 101.2 $kNm^{-2}$ or Pa).

There are two categories of heat; these are:

a. Higher heat of combustion that includes the heat of the water vaporization

b. Lower heat of combustion that does not include water vaporization heat

The **heating value** (or energy value or calorific value), also called higher heating value (HHV) (or gross energy or upper heating value or gross calorific value (GCV) or higher calorific value (HCV)), is determined by bringing all the products of combustion back to the original precombustion temperature, and, in particular, condensing any vapor produced.

Lower heating value (LHV) or net calorific value (NCV) or lower calorific value (LCV) is determined by subtracting the heat of vaporization of the water vapor from the higher heating value, as follows:

$$HCV = LCV + h_v \times (nH_2O_{out}/nfuel_{in})$$

Where: $h_v$ is the heat of vaporization of water,

$nH_2O_{out}$ is the moles of water vaporized and
$nfuel_{in}$ is the number of moles of fuel combusted

It may be assumed that hydrogen in the fuel reacts with oxygen to give water by ratio of one part to eight to form nine parts of water. In fact:

$$H_2 + 1/2\,O_2 \rightarrow H_2O$$

$$2H + 1/2O_2 = H_2O$$

$$2\text{parts} + 16\text{parts} = 18\text{parts}$$

$$1\text{parts} + 8\text{parts} = 9\text{parts}$$

Amount of water produced by burning unit mass of fuel = 9H/ 100 g => Latent heat of steam = 2,458 J/g => then condensation of steam = 9H/100 x 2,458 J

$$\Rightarrow NCV = [HCV - 9H/100 \times 2458] = [HCV - 221H]\ kJ/kg$$

The calorific value can be determined by Dulong formula as follows:

$$\text{Calorific value} = 32{,}851C + 141{,}989((H-O)/8) + 9{,}263S\ (kJ/kg)$$

where: C, H, O, S refer to % (fraction) of carbon, hydrogen, oxygen and sulfur, respectively.

to complete the chemical reaction at the stoichiometric level of oxidation of material. It is a technological complex process that needs several control procedures and operations.

This process passes through three main stages: drying, thermal decomposition (pyrolysis) and complete combustion (burning). These stages usually take place at a temperature around 800–1000 °C. The waste materials are converted into hot gases (mainly $CO_2$, $N_2$, $O_2$, hydrocarbons and $H_2O$), ashes and heat which is used to produce a steam that can be used in turbine generator to produce electricity. For detailed information about combustion reactions and furnaces refer to specialized textbooks [3].

Incineration can be used to treat industrial waste materials (hazardous or nonhazardous, including clinical or hospital waste materials), commercial and municipal solid waste materials. The aims of thermal treatment include volume reduction of waste, reduction of the concentration of waste contamination, recover energy and produce secondary raw material such as ashes and/or gases. Before the thermal treatment, the most important properties of waste are moisture content, during and after treatment calorific value, volatile matter content and ash content.

*Effect of moisture or water content:* It is possible to burn any type of biomass or waste materials. However, the combustion is related to moisture content of these materials. Therefore, waste materials with higher moisture content must be pre-dried before burning. The waste materials that have high moisture content are more adequate to biological processes such as AD.

Industrial waste such as bagasse typically has a range of moisture content between 45 and 55 percent by weight [4,5]. There are several types of bagasse dryers [6,7,8]. Sosa-Arnao et al. (2004) reviewed the bagasse drying system to improve the boiler energetic efficiency [9]. As burned basis, bagasse normally has a heating value between 7 and 9 MJ/kg [5] on a wet basis. This value usually depends on efficiency of the process and the degree of dryness of the material [9]. Box 9.2 explains the procedure for the calculation of combustion heat.

On the other hand, dryer bagasse has a higher calorific value; some heat is required to evaporate its moisture. The high heating value (HHV) is typically between 18 and 20 MJ/kg [10], and approximately 4,500 mg/m$^3$ of fly ash is present in the flue gases from the combustion of bagasse [11]. The typical calorific value of MSW is about 9 MJ/kg [12]. Box 9.3 provides an example on combustion heat calculation.

Incineration technologies have some advantages such as (a) reducing the weight and volume of the waste; (b) minimizing the hazards and contaminants of waste; (c) recovering energy which reduces costs of waste treatment; (d) producing ash materials that could be used in construction such as fills for dams and roads or be mixed with cement. On the other hand, these technologies have some disadvantages such as high capital cost, need for a special technical operation and special management procedures in the case of hazardous waste treatment.

The incineration technologies can be divided into two types: *large-scale incineration technology*, such as mass burn incineration with capacity of about 10–50 tons per hour, and *low-scale incineration technology* with a capacity of 1–2 tons per hour as fluidized bed and rotary kiln incinerators. There are different types of waste incinerators such as rotary kiln, starved air, fluidized bed, multiple-hearth and so on.

Figure 9.2 illustrates a typical incinerator plant of municipal solid waste produced by Hitachi Zosen Inova and located in a former industrial zone north of Barcelona. The capacity of the plant is about 360,000 tons of municipal solid waste per year [13].

**BOX 9.3    EXAMPLE**

To complete combustion of 50 tons of waste materials, calculate the minimum amount of air required and the calorific value. A sample of 100g has been analyzed and found to contain, by weight, 75% C, 8% $O_2$, 5% $H_2$, 3% S and the rest $N_2$.

**Solution**

   a)   Determine the quantity of air
       1.   $N_2$ is not involved in combustion; calculate the total weight of $O_2$ required for combustion of C, $H_2$ and S; then subtract the available $O_2$ in the sample to arrive at the net $O_2$ required. Convert this to the amount of air required by the given weight of the sample.
       2.   $C + O_2 => CO_2$ (12g C required 32g $O_2$)
       3.   75g C will then require $(32 \times 75)/12 = 200g\ O_2$
       4.   Likewise, 5g $H_2$ will require $(16 \times 5)/2 = 40g\ O_2$ and
       5.   3g S will need $(32 \times 3/32) = 3g\ O_2$
       6.   Total $O_2$ required for 100g sample = 200 + 40 + 3 = 243g;
       7.   The sample has 8g $O_2$ => net $O_2$ required = 243 – 8 = 235g; Composition of air: (By weight 23% $O_2$; 77% $N_2$; by volume 21% $O_2$; 79% $N_2$), % by weight of $O_2$ in air is 23%; the quantity of air required by 50 ton of the fuel (waste) will be $(235 \times 50)/23 = 510.9$ tons.
   b)   Using the Dulong formula to determine calorific value:

Calorific value $= 32,851C + 141,989\big((H - O)/8\big) + 9,263S\,(kJ/kg)$

$$= 32,851 \times 75 + 141,989 \times \big((5 - 8)/8\big) + 9,263 \times 3\,(kJ/kg)$$

$$= 2463825 - 53246 + 27789 = 2438.368\,MJ/kg$$

A brief description of the unit functions and processes of incinerator plant is provided in the following paragraphs:

*Receiving of waste and control:* This process takes place in several steps; firstly, batch waste materials are delivered into the plant. Next they will be weighted to determine the total quantity entrance. Next, they will be sorted or screened to separate noncombustible materials (such as metal

| Waste receiving and storage | Combustion and boiler | Flue gas treatment | Energy recovery | Residue handling |
|---|---|---|---|---|
| 1  Tipping hall | 4  Feed hopper | 12 Electrostatic precipitator | 17 Turbine | 20 Wet conveyor |
| 2  Waste pit | 5  Ram feeder | 13 Scrubber | 18 Trafo | 21 Bottom ash extractor |
| 3  Waste crane | 6  Hitachi Zosen Inova grate | 14 Bag filter | 19 Power export | 22 Vibrating conveyor |
|  | 7  Hydraulic station | 15 Induced draft fan |  | 23 Bottom ash bunker |
|  | 8  Primary air fan | 16 Stac |  | 24 Bottom ash conveyor |
|  | 9  Secondary air fan |  |  | 25 Residue silo |
|  | 10 Membrane walls |  |  |  |
|  | 11 Boiler (existing) |  |  |  |

**FIGURE 9.2**

Typical incinerator plant of waste materials. The table explains the parts of the system. (Courtesy of Hitachi Zosen Inova, 2013 (H.Z. Inova, 2013, Waste Plant in City of Sant Adrià de Besòs, Zone North of Barcelona, Spain).

cans and glass), recovery materials such as plastic, paper and metals, and to determine quality of combustible waste material and calorific value. Finally, the results will be registered in a database.

*Storage of waste materials:* After sorting and pretreatment, the combustible waste materials (in order to reduce the size by shearing mills or crushers, homogenization and mixing) will be stored in a place called bunker or hopper system. Then, charging of the waste materials will take place in a continuous feed into the incinerator by overhead crane that is operated through funnel tubes that make the input into the incinerator or the charging mechanism.

*Furnace*: The furnace or grate systems used in the waste incineration are commonly divided into three generic types: fixed furnace, rotary furnace "kiln" and fluidized bed furnace. The selection of the most adequate kind depends on the type and quantity of waste materials. In general, the waste materials are fed into the furnace by gravity (usually downward angle over 25° which depends on the design of furnace). During this move over the grate systems, the materials pass, in the furnace, by three zones or stages, these are drying, combustion and burnout (Figure 9.3). A brief description of each zone is provided in the following paragraphs.

*Waste drying*: In the first stage in furnace, the waste materials will be dried at a temperature more than 100 °C (all water and volatile contents in the waste will be evaporated) by thermal radiation and/or convection heat and/or by preheated air (when waste materials are transported through the grate system). Usually, the furnace needs an auxiliary fuel for preheat and dry of the waste materials before the materials ignition and burning start.

*Waste burning*: After drying, the waste materials pass into burning zones where it will be burned completely if enough air quantities are present. In case the oxygen quantity is less than the stoichiometric values, then the partial oxidation (gasification) will take place. In the zones where

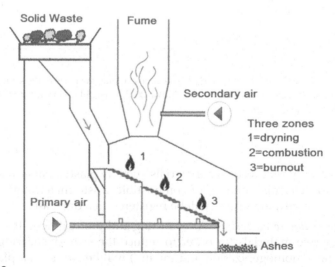

**FIGURE 9.3**

Illustration of incineration process zones.

no oxygen arrives, the materials decompose by heating; this process is known as pyrolysis. On the other hand, at the center of mass, when there is not enough oxygen or when oxygen levels tend to be zero, the thermal decomposition is known as pyrolysis/gasification. Pyrolysis or gasification stage takes place when the temperature is more than 250 °C [14]. The temperature ranges are about 400 to 700 °C in the case of pyrolysis, 700 to 1,000 °C in the case of gasification, and 800 to 1,450 °C in the case of complete combustion [15]. When the temperature reaches 250 °C, the materials decomposition stage will begin; this will cause gases, vapors of water and volatile compounds to be released. In the case of no complete combustion, the process produces potentially hazardous soot and high molecular weight hydrocarbons. To complete the combustion process, waste materials require a secondary supplied air.

After the waste materials pass the combustion zone, the burned-out residue arrives to *burnout zone* that monitors the burnout quality. At this zone, additional residence time will be available to complete the burning and reduce the temperature of the ashes which will be discharged from the bottom of the furnace.

Typically, the temperature of furnace has a range between 750 and 1,000 °C, and reaches up to 1,600 °C. The diagram in Figure 9.4 briefly describes the technological overview of incineration facilities and energy recovery from waste.

*Grate firing system*: The grate is the heart of the combustor; the waste material moves down by gravity and by the action of the grate movement. Depending on the design of combustor (vessel geometry), different design exists such as rocker grate, roller grate, travelling grate or horizontal stoker-type and reciprocating grates [16].

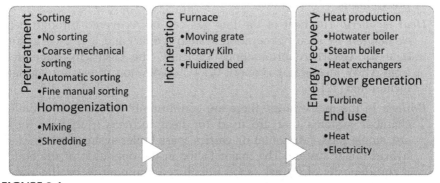

**FIGURE 9.4**

An overview diagram of waste incineration facilities.

Types of incinerator chamber design

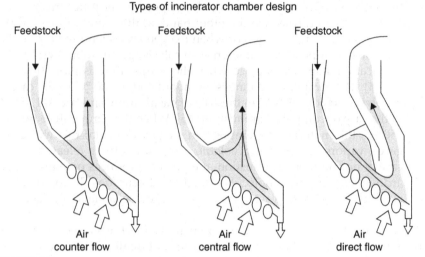

**FIGURE 9.5**

Types of flow circulation in combustion chambers.

*Air circulation:* The flow circulation of air and hot flue gases in the combustion chamber is related to combustion efficiency. The circulation depends on the design of chamber. Different designs exist; these are direct flow, counter flow and central flow (Figure 9.5).

*Slag and ash removal:* From the bottom of the grate and after the waste materials are completely burnt, the rest noncombustible waste materials will be in the form of ash which can be collected as slag and particulate vitrification. Ash residue will be moved and disposed of in landfill sites or they can be used in constructions such as fill for dams and roads or mixed with cement. Fly ash can be recovered from gas clean-up systems as particulate materials by the cyclones, bag filters and electrostatic precipitators.

*Heat recovery:* The heat of the flue gas can be recovered through heat exchangers by steam generation system (boiler) and it is used in turbine for electricity production. Typically, the electricity generated from MSW incinerator is about 0.3–0.7 MWh/ton of waste materials incinerated [12].

*Boilers:* In the incinerators, there are communally three types of boiler chamber systems that are used for heat recovery purposes; these are: single pass, horizontal or vertical pass boiler system or both and hybrid boiler system. The major factor in the operation of the boiler is the deposits of fly ash, soot, volatilized compounds on the water tubes that reduce the efficiency of heat transfer from flue gases to water. To minimize this phenomenon, it is necessary to arrange the tubes in parallel to the gas flow. Blower can clean the deposits of soot. Also, wet cleaning with superheated steam and then treating the wastewater produced can be applied.

Several case studies have been reported in the literature on biomass use technologies. Larson et al. (2001) [17] reviewed a biomass integrated-gasifier/gas turbine combined cycle (BIG/GTCC) technology systems. Broek et al. (1997) [18] described the initiatives of two Nicaraguan sugar mills (San Antonio and Victoria de Julio) to supply electrical power to the national grid; power generation based on dedicated energy crops during the sugarcane harvesting season with bagasse as its principal fuel and outside this season with eucalyptus grown as dedicated energy crop.

*Emission control:* The flue gas contains several gases such as carbon oxides ($CO_x$), nitrogen oxides ($NO_x$), sulfur oxides ($SO_x$), water vapor, polycyclic aromatic hydrocarbons, furans, dioxins, unburned particles and other pollutants. The dusts or particulates can be removed from the flue gas by cyclones, filters and/or electrostatic precipitators. Scrubber can be used to treat the gaseous pollutants. After the treatment, gases will be disposed through stack (dispersion of emissions through the chimney stack). The pollutants emitted to the atmosphere must be under the regulations and legislations of environment. For more detailed information about gas clean-up and treatment methods, the reader is invited to see Chapter 11.

Thermal treatment in the incineration methods include open burning which is usually made without any air pollution control. The method is used to treat MSW by burning piles of waste outdoor or burning them in a burn barrel in several urban centers to reduce the volume of waste received at the dump (open dump). The open burning could have serious effects on human health and environment (see Chapter 2).

For detailed information about incineration technologies, incinerators types and incineration methods, the reader is referred to specialized literature such as Georgieva and Varma (1999) [19], Tchobanoglous and Kreith (2002) [20], Trinks et al. (2004) [21]; Williams (2005) [12] and Mullinger and Jenkins (2014) [22].

---

## 9.3 Pyrolysis

Pyrolysis is a thermal process and can be applied for treatment of organic waste (biomass) destruction in the absence of oxygen to produce liquid (termed bio-oil or bio-crude), solid (charcoal) and gaseous fractions (syngas). The process condition depends on the end product desired. There are slow and fast pyrolysis process [23]; this could be either the production of fuel gases or the complete destruction of the waste materials. The complete elimination of air from waste materials is very difficult in practice; therefore, the pyrolysis process is usually performed in a neutral atmosphere of argon, helium or nitrogen which acts as carrier flow [24,25].

The pyrolysis mechanism can be explained as follows: waste materials contain different types of submaterials such as papers and plastics. These materials have complex polymers (chemical compounds) which will be degraded under the action of heat (thermal pyrolysis without oxygen) into molecules of less molecular weight. In fact, the breaking down of the natural organic molecular chains, such as cellulose, hemicellulose and lignin, and/or synthesis of large polymer (plastics) produce gases, liquid (tar) and solid, the so-called char. The tar is a black mixture of free carbon and hydrocarbons as polycyclic aromatic hydrocarbons. The gases are condensable and non-condensable. The condensable gases will have the characteristic of a liquid or oil, while non-condensable are gases. The final product depends on the temperature, feedstock, heating rates and the time of process.

Usually, the process temperatures are in the range of 400–900 °C. Short exposure to high temperatures is termed "Flash pyrolysis", which maximizes the amount of liquids generated, and is normally operated at around 500–700 °C. If lower temperatures are applied for longer periods of time under vacuum, chars predominate compared to fast pyrolysis product, and the process is called "Vacuum pyrolysis".

The final products can have several uses; in the case of waste feedstock, it is usually used as a fuel for energy production. Pyrolysis oils can be used as raw materials for specialty chemicals production and/or liquid fuel (fuel oil). The pyrolysis solid char can be used as a solid fuel and/or as an activated carbon. Pyrolysis gases can be used as fuel for the pyrolysis process itself and/or as fuel gas that is rich with hydrogen. Pyrolysis gas contains carbon monoxide ($CO$), carbon dioxide ($CO_2$), hydrogen ($H_2$) and hydrocarbon compounds such $CH_4$, $C_2H_8$ and traces compounds as ammonia ($NH_3$), hydrogen sulfide ($H_2S$) and hydrogen chloride ($HCl$).

Rotary kilns, fluidized beds, fixed-bed reactors and other reactors are technologies used for waste pyrolysis treatment. Figure 9.6 provides a schematic diagram of the pyrolysis process of waste.

## 9.4 Gasification

Like pyrolysis, gasification is the thermo-chemical reduction of a waste based upon partial oxidation (limited oxygen supply) without direct combustion.

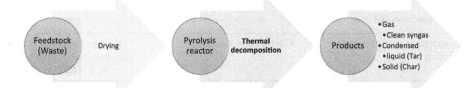

**FIGURE 9.6**

Schematic diagram of the pyrolysis process of waste.

The waste materials react with agents, usually air and/or steam under the application of heat. Process temperatures are in the range of 500–1,500 °C. The main product is typically a gas that is called "Syngas" and solid known as char. Syngas is a mixture of gases; these contain methane ($CH_4$), carbon monoxide ($CO$), carbon dioxide ($CO_2$), hydrogen, nitrogen ($H_2$), water vapor ($H_2O$), nitrogen ($N_2$), few amounts of higher hydrocarbons as ($C_nH_m$) and inorganic impurities and particulates as alkali metals, HCN, $NH_3$, $H_2S$ and HCl. The syngas produced from waste of sugarcane industry does not cause environmental problems during the production because contamination of sulfur and nitrogen are quasi-negligible [5].

There are several utilizations of the syngas product; it can be burned to create heat and steam to produce electricity. When syngas is directly used in internal combustion engines, it needs higher cleaning, while in other cases it needs less treatment. It can also be converted into methanol, ethanol and other chemicals or liquid fuels. The solid residues (char) can be burned to provide heat for the gasifier reactor itself or to produce steam. The typical waste gasification reactions that take place at sub-stoichiometric conditions are listed in Table 9.1.

The Boudouard (gasification of char in carbon dioxide) and steam reactions are endothermic, while the oxidation reactions are exothermic. The gasification reactions depend on the gasification agents (oxygen, air, steam, hydrogen, carbon dioxide), and the carbon contained in the materials of waste. Detailed information on the sequential steps of endothermic and exothermic reactions related to gasification process are widely available in the literature, e.g. Knoef (2005) [26].

Fluidized bed gasifier, rotary kiln gasifier, entrained flow gasifier, updraft and downdraft gasifier and other reactors designs are technologies used for waste gasification treatment. For detailed information about these gasifiers' designs, the reader is advised to refer to specialized publications such as Prabir Basu (2006) [27] for updraft gasifier; Prabir Basu (2010) [28] for downdraft gasifier; Prabir Basu (2013) [29] for bubbling fluidized-bed gasifier. Further details on gasification can be found in Souza-Santos (2004) [30].

**TABLE 9.1**

Main reactions of gasification

| Gasification agents or type reaction | Gasification reactions | Heat of Reaction (kJ/mol, at T = 1000 K, P = Po)[a] |
|---|---|---|
| Oxygen | $C + 1/2O_2 \rightarrow CO$ | -112 |
| | $CO + 1/2O_2 \rightarrow CO_2$ | -283 |
| | $H_2 + 1/2O_2 \rightarrow H_2O$ | -248 |
| Steam | $C + H_2O \rightarrow CO + H_2$ | 136 |
| | $CO + H_2O \rightarrow CO_2 + H_2$ | -35 |
| | $CH_4 + H_2O \rightarrow CO + 3H_2$ | 206 |
| Boudouard reaction | $C + CO_2 \rightarrow 2CO$ | 171 |
| Hydrogenation reactions or methanation reaction | $C + 2H_2 \rightarrow CH_4$ | -74.8 |
| | $CO + 3H_2 \rightarrow CH_4 + H_2O$ | -225 |
| | $CO2 + 4H_2 \rightarrow CH_4 + 2H_2O$ | -190 |

[a] P. Kannan, A. Al Shoaibi and C. Srinivasakannan, 2012, "Optimization of Waste Plastics Gasification Process Using Aspen-Plus" in Y. Yun (ed.) "Gasification for Practical Applications", InTech, Croatia, ISBN: 978-953-51-0818-4.

Finally, to be emphasized is that the thermal treatment processes of waste can be combined as gasification and pyrolysis or gasification and combustion or vice versa or all three processes.

# Review Questions

What is the difference between pyrolysis and gasification?
What are the main technologies of incineration?

# References

1. Burning Bagasse, available online at http://energyconcepts.tripod.com/energyconcepts/bagasse.htm Burning Bagasse.
2. S. Arni and A. Converti, 2012, "Conversion of Sugarcane Bagasse into a Resource" in João F. Goncalves and Kauê D. Correia (eds.) "Sugarcane: Production, Cultivation and Uses", Series: Agriculture Issues and Policies, 1st Quarter, ISBN: 978-1-61942-214-8.
3. P. Mullinger and B. Jenkins, 2008, "Industrial and Process Furnaces Principles, Design and Operation", First edition, Elsevier, Oxford, UK.

4. W. Alonso-Pippo, C.A. Luengo, F.F. Felfli, P. Garzone and G. Cornacchia, 2009, "Energy Recovery from Sugarcane Biomass Residues: Challenges and Opportunities of Bio-oil Production in the Light of Second Generation Biofuels", Journal of Renewable Sustainable Energy, Vol. 1, p. 063102.

5. D. Janghathaikul and S.H. Gheewala, 2006, "Bagasse—A Sustainable Energy Resource from Sugar Mills", Asian Journal of Energy Environment, Vol. 7, Supplement 03, pp. 356–366, ISSN: 1513–4121, available online at www.asian-energy-journal.info.

6. V.J. Bailliet, 1976, "Bagasse Drying versus Air Pre-heating", The Sugar Journal, Vol. 38 (10), pp. 52–53.

7. A. Arrascaeta and P. Friedman, 1984, "Bagasse Drying: Past, Present and Future", International Sugar Journal, Vol. 86, Supplement 1021, pp. 3–6.

8. A. Arrascaeta and P. Friedman, 1987, "Bagasse Drying", International Sugar Journal, Vol. 89, Supplement 1060, pp. 68–70.

9. J.H. Sosa-Arnao, F.M. de Oliveira, J.L.G. Corrêa, M.A. Silva and S.A. Nebra, 2004, "Sugarcane Bagasse Drying – a Review", Drying 2004 – Proceedings of the 14th International Drying Symposium (IDS 2004), São Paulo, Brazil, vol. B, pp. 990–997.

10. Janghathaikul and Gheewala, 2006, "Bagasse – A Sustainable Energy Resource from Sugar Mills".

11. Sosa-Arnao, de Oliveira, Corrêa, Silva and Nebra, "Sugarcane Bagasse Drying – a Review".

12. S. Arni, B. Bosio and E. Arato, 2010, "Syngas from Sugarcane Pyrolysis: An Experimental Study for Fuel Cell Applications" Renewable Energy, Vol. 35, pp. 29–23, available online at: http://dx.doi.org/10.1016/j.renene.2009.07.005.

13. Multilateral Investment Guarantee Agency Environmental Guidelines for Sugar Manufacturing, page 495.

14. P.T. Williams, 2005, "Waste Treatment and Disposal", 2nd Edition, Wiley, ISBN 0-470-84912-6.

15. H.Z. Inova, 2013, Waste Plant in City of Sant Adrià de Besòs, Zone North of Barcelona, Spain. www.hz-inova.com; www.hz-inova.com/cms/wp-content/uploads/2014/11/hzi_referenz_barcelona_en.pdf

16. M. Juma, Z. Koreňová, J. Markoš, J. Annus and Ľ. Jelemenský, 2006, "Pyrolysis and Combustion of Scrape Tire", Ptoleum & Coal. ISSN: 1337–7027.

17. D.'H. Hassan, 2014, "Best Available Techniques Reference Document on Waste Incinerator" Department of the Environment, Malaysia. File in pdf format available on website: www.doe.gov.my/portalv1/wp-content/uploads/2014/07/BEST-AVAILABLE-TECHNIQUES-GUIDANCE-DOCUMENT-ON-WASTE-INCINERATOR.pdf

18. C.R. Brunner, 2002, "Incineration Technologies", in George Tchobanoglous and Frank Kreith (eds.), "Handbook of Solid Waste Management", 2nd Edition, McGraw-Hill.

19. Williams, "Waste Treatment and Disposal".

20. E.D. Larson, R.H. Williams and M. Regis L.V. Leal, 2001, "A Review of Biomass Integrated-Gasifier/Gas Turbine Combined Cycle Technology and its Application in Sugarcane Industries, with an Analysis for Cuba", Energy for Sustainable Development, Vol. 5 No. 1, March 2001.

21. R. van den Broek, R.C. de Miranda and A. van Wijk, 1997, "Combined Heat and Power Generation from Bagasse and Eucalyptus by Sugarmills in Nicaragua", Article presented at the Third Biomass Conference of the Americas, 24–29 August 1997, Montreal.

22. K. Georgieva and K. Varma, 1999, The World Bank Technical Guidance Report, "Municipal Solid Waste Incineration", Washington, United States of America.

23. G. Tchobanoglous and F. Kreith (eds.), 2002, "Handbook of Solid Waste Managament", 2nd Edition, McGraw-Hill Companies, 0-07-150034-0, 0-07-135623-1.

24. W. Trinks, M.H. Mawhinney, R.A. Shannon, R.J. Reed and J.R. Garvey, 2004, "Industrial Furnaces", 6th Edition, Wiley. ISBN 0-471-38706-1.

25. Williams, "Waste Treatment and Disposal".

26. P. Mullinger and B. Jenkins, 2014, "Industrial and Process Furnaces: Principles, Design and Operation", 2nd Edition, Elsevier Ltd. ISBN-13: 978-0-08-099377-5.

27. S. Arni, 2018, "Comparison of Slow and Fast Pyrolysis for Converting Biomass into fuel", Journal of Renewable Energy, Vol. 124, pp. 197–201, available online at http://dx.doi.org/10.1016/j.renene.2017.04.060

28. S. Arni, 2004, "An Experimental Investigation for Gaseous Products from Sugarcane by Fast Pyrolysis", Energy Education Science and Technology, Vol. 13, Supplement 2, pp. 89–96.

29. S. Arni, 2004, "Hydrogen-Rich Gas Production from Biomass via Thermochemical Pathways", Energy Education Science and Technology, Vol. 13, Supplement 1, pp. 47–54.

30. Janghathaikul and Gheewala, 2006, "Bagasse – A Sustainable Energy Resource from Sugar Mills".

31. H. Knoef, 2005, "Practical Aspects of Biomass Gasification" in H. Knoef (ed.) "Handbook of Biomass Gasification", BTG-Biomass Technology Group (BTG), Enschede, the Netherlands.

32. P. Basu, 2006, "Combustion and Gasification in Fluodizid Beds", Taylor & Francis Group, LLC.

33. P. Basu, 2010, "Biomass Gasification and Pyrolysis Practical Design and Theory", Elsevier Inc. ISBN: 978-0-12-374988-8.

34. P. Basu, 2013, "Biomass Gasification, Pyrolysis and Torrefaction – Practical Design and Theory", Second Edition, Elsevier Inc. ISBN: 978-0-12-396488-5.

35. M.L. de Souza-Santos, 2004, "Solid Fuels Combustion and Gasification: Modeling, Simulation, and Equipment Operation", Marcel Dekker, Inc. ISBN: 0-8247-0971-3.

# 10

## Industrial Wastewater Treatment

*Key Learning Objectives*

- Understanding the wastewater contaminants.
- Understanding the principal steps of wastewater treatment.

## 10.1 Introduction

This chapter is about water contamination treatment. Human activities including domestic, agriculture and industrial sectors continuously produce wastewater. Wastewater is a contaminated water that is produced after being used in homes, public institutions, commercial establishments, industries and similar entities. Sometimes it is called *sewage*.

Domestic wastewater can be derived from toilets (known as *black water*) or from other domestic activities (known as *gray water*) [1]. The domestic wastewater that is contaminated with human wastes is usually called *sanitary* wastewater, while industrial wastewater is called *non-sanitary* wastewater. Most pollutants of water come from industrial activities. Typical examples are water used in processes such as industrial production, reactants, solvents, absorption, energy transport (heat exchange as cooling), washing and rinsing. After being used in processes, water contaminated with components that deteriorate its quality in any way must be treated before reusing or discharging in a sewer system or into receiving water bodies.

In the industrial sectors, the *on-site treatment* is done by the owner, while in the urban sectors the *off-site treatment* is done by the municipal authority.

Disposal of industrial wastewater requires pretreatment before it is discharged into a municipal sewer system. In general, the pretreated wastewater from industrial sites and households is connected to public sewer systems. However, industrial wastewater discharged from manufacturing processes must comply with the applicable regulations.

The Presidency of Meteorology and Environment (PME) established the standards for wastewater discharge in the KSA (see the standards and regulation on the PME website). Depending on the type of manufacturing activities and processes, the owner is responsible for the treatment of the wastewater according to the environmental regulations. In any way, the industrial wastewater must be minimized, diluted or treated, before disposed of.

There are several treatment technologies that can be used to remove contaminants or recover water [2]; these include centrifugation, filtration, ultrafiltration, reverse osmosis, chemical coagulation and flotation processes. Treatment processes can include physical, chemical and biological (see Chapter 7). In fact, wastewater treatment systems consist of a combination of physical, chemical and biological separation processes [3]; the physical–chemical separation is, usually, conducted as pre- and post-treatment of the biological separation processes. The sequential steps of treatment operations, in order to increase treatment level, are the preliminary, primary, secondary and tertiary (advanced) treatment steps. These operations have the scope of solid/liquid separation to remove solids, organic matter and disinfection [4]. Physical treatment includes sedimentation, filtration, flotation, stripping, adsorption, ion exchange and other processes. Chemical treatment includes chemical oxidation and reduction, chemical precipitation and other chemical treatment. Biological treatment includes aerobic and anaerobic.

Selection of the adequate treatment technologies depends on water chemical, physical and biological proprieties; specifically, total daily quantity of flow rate (F), biochemical oxygen demand (BOD), chemical oxygen demand (COD), total dissolved solids (TDS), total suspended solids (TSS), fats, oils and greases (FOG), heavy metal and chemical ions dissolved, acidic and basic characteristics, toxic and nontoxic characteristics.

The quality of wastewater treatment depends on its specific end reuse purposes. The quality parameters required for safe use of recycled wastewater include turbidity, dissolved minerals, specific inorganic and organic contaminants (as ammonia, phosphorus, etc.), residual organics and disinfection for microbial pathogens. Water recycling in the industrial sector (Figure 10.1) provides major benefits to industry in terms of reducing the quantity of fresh water supply, reducing the generated wastewater and reducing the total net consumed water and costs.

## 10.2 Types of Water Contaminants

Before dealing with wastewater treatment methods, it is necessary to describe the major types of water contaminants.

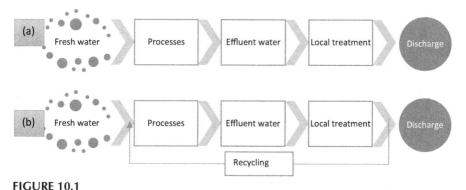

**FIGURE 10.1**

Using water in the industrial sectors (a) without recycling, (b) with recycling.

Classification of water pollutants, based on the nature and source of contaminant, can be made into organic, inorganic and biological pollutants. Organic pollutants mainly include fats, oil and grease (FOG); these are produced from food processing wastes and petrochemical industrials. Volatile organic compounds (VOCs) are produced from industrial solvents and petrochemical industrials. In addition, other organic compounds are produced from several industrial discharges such as petroleum hydrocarbons and pharmaceuticals industries.

Inorganic pollutants mainly include heavy metals produced from metallurgic industrials. Ammonia, nitrates and phosphates are also produced from industrial fertilizers and food processing waste. Sulfur dioxide is produced from power plants while other chemical wastes are produced from several industrial discharges. Biological pollutants mainly include bacteria, viruses and specific pathogens that come from different sources such as toilets.

Furthermore, wastewater has specific physical, chemical and microbiological characteristics [5]. Figure 10.2 provides a summary of these characteristics.

*Physical characteristics*: These include solid contents, turbidity, color, taste, odor and temperature. Solids contents can be found as dissolved; settleable (residues after filtration) or suspended in water that caused the turbidity. These could have an organic or an inorganic origin.

Turbidity is a suspension of solid matters in water that makes the water less visible or changes its color by the perception of different chemical compounds. It also causes change of water taste and odor based on the concentration of the contaminants.

Colors are typically derived by vegetable decomposition, dyes or metallic ions; for example, blue color is caused by copper, green by nickel, yellowish brown by iron, yellowish green by chrome [6,7].

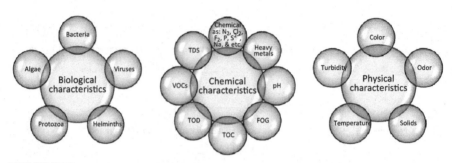

**FIGURE 10.2**

Summary of wastewater physical, chemical and biological characteristics.

Gases derived from the decomposition of materials and substances that are present in water usually cause wastewater odor. For example, the odor of ammonia is due to contamination with ammonia ($NH_3$); the odor of rotten eggs is caused by contamination with hydrogen sulfide ($H_2S$); the odor of rotten cabbage is caused by contamination with dimethyl sulfide (($CH_3)_2S$), or diphenyl sulfide(($C_6H_5)_2S$); odor of fish is caused by contamination with methyl amine ($CH_3NH_2$)[8]. The typical color of municipal wastewater is usually gray.

Variation in temperature of water has an effect on the variation of concentration of the chemical compounds in water, viscosity, pressure and thus on the efficiency of treatment.

In addition, the flow rate ($m^3$/day) is considered as a physical characteristic. *Chemical characteristics*: Chemical characteristics of wastewater include toxicity and its contents of elements, compounds and ions. These can be associated with organic and inorganic contents such as:

a) Total oxygen demand (TOD) which consists of both the biochemical oxygen demand (BOD) and the chemical oxygen demand (COD). BOD and COD are, respectively, a measure of biodegradable and oxidizable matter in the wastewater (see the biological processes, Chapter 7), volatile organic compounds (VOCs) and total organic carbon (TOC).

b) Heavy metals (as chromium, cobalt, copper, iron, lead, mercury, manganese, zinc, etc.), nitrogen (ammonia, nitrite and nitrate), sulfur ions (sulfur dioxide, sulfate, sulfide), chlorides, fluorides, phosphorus, sodium and others.

In addition, these compounds can change water properties such as the water acidity and alkalinity (pH), salinity and hardness. Also, fats, oils and grease (FOG) are considered chemical characteristics of water.

*Microbiological characteristics*: Typically, influent wastewater contains numerous organisms; the majority of these are nonpathogenic and many are pathogenic. The microbiological characteristics of wastewater includes pathogenic organisms like bacteria (as *Escherichia coli, Salmonella*, etc.), viruses (as *Adenovirus, Enteroviruses*, etc.), protozoa (*Balantidium coli, Giardia lamblia*, etc.), algae, helminths [9]. Biological characteristics help in the water treatment (biological treatment). In general, physical, chemical and biological qualities of water are very important factors in wastewater treatment.

The effects of wastewater contaminants depend on source, type and concentration of pollutants. These effects leave their impact on the human health, animals (as toxicity effects) and the environment (see Chapter 2). The main scope of treatment is to protect public health, water supplies and the environment.

## 10.3 Wastewater in the Industrial Sector

The treatment techniques of wastewater depend on the type of industrial pollutants, concentrations and their classifications. Usually, the industrial wastewater includes toxic materials, chemicals, acids, alkalis, dyes, grit, oil, waxes and detergents.

The main wastewater sources in the industrial sector are:

a) Washing: Washing and cleaning of raw materials and equipment as backwashing of filters, ion exchangers and gas clean-up systems;
b) Production processes: Such as food production and related processes, metals production and their processes and chemical syntheses;
c) Waste treatment: Such as landfill leachates;
d) Utilities: Such as cooling cycles and boiler systems;
e) Others: Inflow and rainwater.

Usually, an industrial plant has many processes and utilities; each one has its wastewater stream with quantitatively, qualitatively and different types of pollution levels [10]. All of these are related to the type and size of industry. In fact, each stream of effluent has a different type of wastewater characteristics; these are: flow rate, pH, color, odor, density, viscosity, COD, BOD, TSS, TDS, ammonia or total nitrogen, total phosphorus, metals present, oil content, toxic or nontoxic elements or any other compounds. It is preferred to treat each stream separately before connecting or mixing them in one process for treatment.

In conformity with the environmental laws and standards (see regulation for water and wastewater), the owner, before conducting any control or treatment implementations, should identify the following effluent characteristics:

a) quantity (flow rate), quality (pollutant characteristics), treatment level (treatment efficiency) and treatment processes (techniques and methods);
b) to examine the wastewater generated during the industrial activity suitable for reusing or recycling;
c) to obtain an effluent discharging permit according to the standards and environment laws.

Treatment techniques in the industrial sector are similar to those applied in the domestic or municipal wastewater treatment, except that they are smaller in size in case of domestic treatment. In general, any industry has appropriate techniques or methods of treatment based on the type and concentration of contaminants. The techniques are especially developed based on the requirements of the site and the type of process. The methods include the separation of soluble and insoluble substances physically or chemically using concentration, precipitation, adsorption, flocculation, acids and bases, redox reaction, ion exchange, reverse osmosis, degassing or stripping and biological methods.

The selection of the most appropriate treatment method or technique depends on the pollutants size range and type of pollutants (Mels and Teerikangas, 2002 [11]; Tchobanoglous et al., 2003 [9]; Takashi Asano et al., 2007 [12], Kemp, 2008 [13]). Wastewater treatment methods in the industrial sector are:

a) Dilution of wastewater in large tanks; the process must be controlled according to the environmental standards before discharging is made.
b) Using chemical–physical and biological methods.

In case that industrial wastewater is connected with the municipal wastewater, it must be pretreated or diluted before the connection occurs in such a way that the operating treatment plant is not affected.

## 10.4 Wastewater Treatment Processes

Wastewater treatment aims to achieve several goals; protect the public health from risks; make water suitable for specific uses such as irrigation or industrial purposes and to safeguard the environment by restoring the natural characteristics of water.

Wastewater treatment can be conducted in three ways; these are batch treatment, semicontinuous treatment and continuous treatment. Batch treatment is a process that is widely applicable in most small and medium sized industries, which do not produce large quantities of wastewater daily. The logical mode is to store the wastewater and then treat them in batches. The semibatch or semicontinuous treatment of wastewater is between batch and continuous, but these are rarely used. The municipal wastewater treatment is a typical use of continuous treatment.

### 10.4.1 Treatment Systems

A domestic or urban wastewater treatment system consists of a series of separation steps or liquid–solid treatment processes (Levine and Asano, 2002 [14]; Asano et al., 2007). The separation steps are known as centralization or off-site. Figure 10.3 illustrates the steps of treatment. There are two principal lines of wastewater treatment; one is used for liquids and the other is dedicated for solids treatment.

In the liquid separation line, Figure 10.3, the steps of treatment can be classified as follows:

*Preliminary treatment:* This aims to remove gross solid materials in order to protect the plant equipment and facilitate the following treatment steps.

*Primary treatment:* This removes the settleable solids and floatable materials (such as FOG).

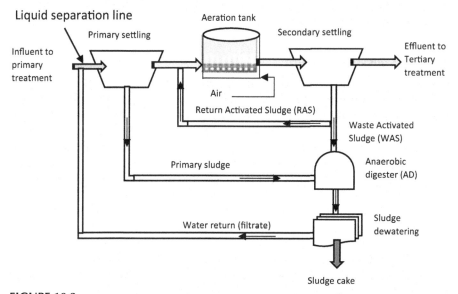

**FIGURE 10.3**

An illustration scheme of solid–liquid separation process in a wastewater treatment system.

*Secondary treatment*: This removes the dissolved and suspended organic matters (such as BOD and TSS) by biological processes. This treatment stage uses separation of activated sludge through aerated reactors, oxidation or stabilization ponds, trickling filters, rotating biological disks, rapid filtration systems and other different systems.

*Tertiary treatment and advanced treatment*: Additional treatment is usually used in conjunction with physical, chemical and biological processes to remove microelements, ions or compounds. These are oxidation, coagulation, flocculation, nitrification and denitrification processes, phosphorus precipitation processes, sedimentation and disinfection. Disinfection is used to kill or remove microorganisms that cause diseases before discharging the water; this is usually done by using chlorination with $Cl_2$ or $NaOCl$.

The second line of wastewater treatment is dedicated to the treatment of solids. This includes:

*Sludge treatment*: This is used to stabilize the sludge (solids removed from wastewater during the treatment) from its activated form into an inactivated form by drying, reducing volume and stabling the pathogenic organisms.

### 10.4.2 Processes of a Treatment System

A wastewater treatment system consists of several processes. These are explained in the following subsections:

#### 10.4.2.1 Collection Systems

Collection system is a network of piping or canals that conveys the wastewater from sources (individual residence as homes and hotels, schools, shopping centers, restaurants, offices, etc.) to treatment plant. The wastewater also includes rainstorms, snowmelt and runoff that enter the public discharging network. These contain amounts of grit, gravel, sand and street debris. The sewer pipes connect several points that start from the small pipes to larger mains, collectors, trunk lines and interceptor, which ends with the treatment plant (Figure 10.4).

The transport is usually made by gravity force, but sometimes when it is not enough to move wastewater through the connection system, pumps (lift stations) are used. The lift stations are considered as an additional cost to the municipal authority (installation, maintenance and operation). In the design, all related parameters, such as minimum and maximum loads of flow rates, must be taken into consideration to ensure adequate collection system, maintenance and capacity of treatment plant.

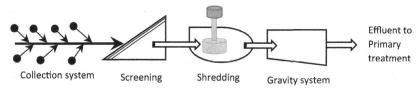

**FIGURE 10.4**

Illustration diagram of collection process and preliminary treatment.

### 10.4.2.2 Preliminary Treatment

Influent enters into treatment plant from the collection system (Figure 10.4). The first stage of treatment is the preliminary treatment. This stage exists in most wastewater treatment plants. It includes several liquid–solid *screening* processes wherein each process, specific materials are removed. For example, screening removes large floating objects, such as plastic bottles, plastic bags, rags, cans, sticks and tree branches, leaves, roots and rocks. Screening is the first unit in wastewater treatment.

Screens have varied forms; coarse screens can be constructed with mesh screens or with parallel bars of steel (bar screen) with openings usually between 5 and 15 cm and placed in an inclined manner across a chamber or channel. The solids can be removed from the screens manually or mechanically then treated as solid waste.

Certain plants use *shredding* devices (as comminutors or barminutors) to reduce the size of solids by catch and then cut or shred, then the grinding materials are later removed in settling phase [15]. The shredding is an alternative way that combines the functions of a screen and a grinder.

After screening is completed, a gravity system is used to remove the gritty materials. Grit removal is a device used to remove gritty materials such as grit and small stones, sand, silt, coffee grounds, cinders, eggshells and other inert materials by settling to the bottom of a grit chamber or tank.

Sometimes, a centrifugal chamber is used to remove grit materials. If the case grist and sand entering the treatment plant cause serious operating problems to the pumps and other equipment, then they need to be collected and removed periodically from the grit chamber. The cleaning system can be manual or mechanical. In certain plants, another screen is added after the grit chamber to provide additional equipment protection. The *gravity chamber* can control the velocity of the wastewater and make some aeration that makes the organic materials stay suspended. Sometimes, additional *sedimentation tank* is added where gravity settling occurs; this increases the efficiency of the separation process.

*Flow measurement* is used to determine water quantity and the organic load entering the treatment plant (for monitoring station). This is necessary for treatment processes and the compliance report data of the plant. Different flowmeters are available to be used for flow measurement such as Magnetic flowmeters and Coriolis Mass flowmeters.

Knowledge of water or wastewater quantity, the type of contaminants and their concentrations are very important factors in the design and treatment operation. For example, the domestic wastewater characteristics continuously change during time. The quantity of wastewater is determined based on probability concepts such as probability distribution analysis. Flow rate average is usually estimated hourly, daily, weekly, monthly or yearly. To control hydraulic overloads and protect equipment (as clogging filters), an equalization device can be installed in the phase of pretreatment.

*Flow equalization* is a technique used to determine the operation parameters of receiving stream of water as flow rate, pollutant levels and temperature usually over all the time (daily). It is a technique to improve successive wastewater treatment processes.

*Preaeration* is an aeration that takes place before primary treatment to facilitate grease removal by floatation, to reduce odors and corrosion by stripping off hydrogen sulfide, to freshen septic waste by air bubbles then reduces $BOD_5$ and to improve solids separation and settling by agitating the solids. Usually, preaeration uses skimming tank or flotation unit.

In certain cases of the pretreatment, chemical substances are added to solve some problems such as neutralizing acids or bases, reducing odors, reducing corrosion, helping grease removal, reducing $BOD_5$ and improving solids removal. Depending on the desired outcomes, the chemical addition can be acids or bases, mineral salts or peroxide, and it can be added in powder or in solution. When a chemical is added, it must be perfectly mixed to ensure successful chemical pretreatment.

### 10.4.2.3 Primary Treatment

The influent of wastewater after treatment in the preliminary treatment unit still contains contaminants as suspended and floatable organic solids. It passes into the primary treatment unit, which is composed of screening and sedimentation tanks (clarification or sedimentation tank), where physical separation process takes place (Figure 10.5). This removes the majority of settleable solids, with a removing range between 90 and 95%, total suspended solids (TSS) between 40 and 60% and the $BOD_5$ in the range 25 to 35% (Drinan and Whiting, 2001 [16]). Only in the settling phase in preliminary and primary stages of treatment, about 75% of the bacteria can be removed or destroyed (Shieh and Nguyen, 1999 [17]).

The primary sedimentor (sedimentation tank or clarifier) has a long rectangular tank shape (basin), a circular tank or other forms. In large plants with big circular tanks, a combination of aeration, mixing, skimming and settling processes take place. Usually, the detention time in a sedimentation tank ranges from 1 to 3 hours (Drinan and Whiting, 2001). FOG or scum and other floating materials can be removed from surface and the

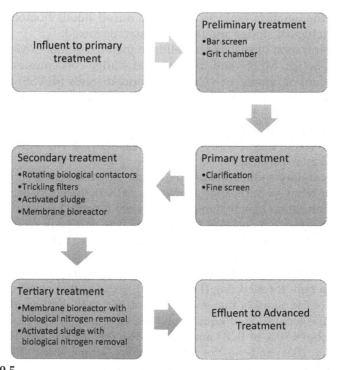

**FIGURE 10.5**

Typical principal steps of wastewater treatment.

heavier settled solids are transported into the bottom by gravity then removed as sludge (so-called primary sludge) by the pump of the sludge unit of treatment.

After the separation of large debris, grit, sand, scum and settleable materials in the preliminary treatment and primary sedimentation, the water stream will have a clear gray color (so-called primary effluent) but still contains dissolved organic and inorganic materials (suspended solids). It is then sent into the secondary treatment phase.

The primary effluent enters into the secondary treatment phase (biological processes, see Chapter 7); this represents the heart of municipal wastewater treatment. In fact, the effluent from the primary clarification of wastewater is delivered into the aeration tank, where inflated air, usually from the bottom of tank, helps the growth of aerobic microorganisms (produce new cells and production) and degrades the organic matters into carbon dioxide, water and other products.

The diffused (aeration) or mechanical aeration and mechanical agitation systems maintain the aerobic environment of the reactor completely mixed (known as mixed liquor, ML). The uniform aeration throughout the reactor (tank) helps the suspended and recycled solids (portion of settled sludge returns to the aeration tank) to make the mixed liquor homogenous and maintains activated sludge and floating materials mixed.

The concentration of suspended solids in the mixed liquor (Mixed Liquor Suspended Solids, MLSS) and the activated sludge concentration in mixed liquor (amount of organic or volatile suspended solids MLVSS) are usually measured in milligrams per liter (mg/l). After the treatment of wastewater in the aeration tank, the mixed liquor passes into the secondary clarifier (sedimentation tank), where solid–liquid separation process takes place by sedimentation (the solid is known as activated sludge). For more details, see Section 10.5 on mass balance.

In the secondary sedimentation, the biodegradable organic matter (accumulation of biomass or biosolid, known as activated-sludge, because it has a high concentration of active microorganisms) is settled by gravity into the bottom of contactor or reactor. Periodically, it is necessary to discharge or remove the sludge in part (Waste Activated Sludge, WAS) into anaerobic treatment unit (known as an aerobic digestion).

Another part of the sludge (Return Activated Sludge, RAS) goes back to the aeration tank for recycling (Figure 10.3). Usually, the recycled portion is about 25% of the quantity that enters the aeration tank (Reuter, 1999 [18]). In this stage, the separation of liquid–solid removes most of the biodegradable materials; about 90% of BOD and 90% of TSS (Sebastian, 1999 [19]; Drinan and Whiting, 2001) and most of the pathogens (Drinan and Whiting, 2001). The clarified effluent produced from the secondary treatment goes into further treatment (advanced treatment).

### 10.4.2.4 Tertiary Treatment and Advanced Treatment

The tertiary and advanced treatment is applied to the water and wastewater to remove pollutants (Figure 10.6) such as soluble COD, heavy metals, phosphorus and nitrogen to improve the quality of effluent to meet discharge environment standards to replenish into water receiving body or groundwater or reuse criteria. The tertiary treatment is usually the final treatment for municipal wastewater. It is used to remove additional suspended solids or organic compounds as TSS and BOD to render the water suitable to be discharged into water bodies, or to be reused in advanced treatment processes.

There are different levels of advanced treatment; the selected advanced level is based on the purpose of the final use, e.g. landscape irrigation, or the type of pollutants needed to remove. Certain physical–chemical separation techniques can be used; these include adsorption, filtration by advanced membranes, reverse osmosis, flocculation and precipitation. Figure 10.6 provides a brief description of the techniques that are used to reduce TSS, BOD, phosphorous and nitrogen compounds.

*Treatment by separation techniques:* Various separation methods can be used in purification of water from micro-contaminants and impurities such as heavy

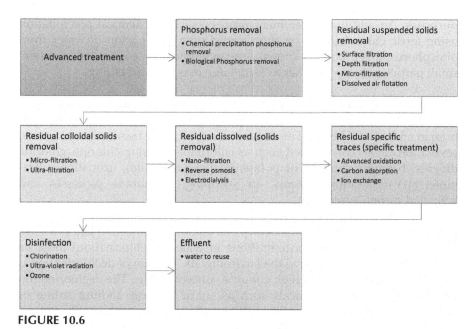

**FIGURE 10.6**

Typical steps of an advanced treatment of wastewater.

metals (arsenic, lead, mercury, radium, iron, manganese), and other chemical compounds such as asbestos, atrazine, fluoride, nitrate, radon, benzene, trichloroethylene, trihalomethanes and silicon. Membrane separation can also be used (as a reverse osmosis) for water reclamation purposes. In addition, water can be demineralized by electrodialysis process.

*Nitrogen removal:* Nitrogen is present in water (wastewater) as an organic nitrogen, ammonia or nitrate compound. Removal of a nitrogen compound, including nutrient, is necessary for several reasons such as to help control algal bloom in the receiving body of water. In addition, the nitrogen, in the form of ammonia, is toxic to fish.

Nitrogen can be removed and controlled by nitrification/denitrification processes (biological processes) or by ammonia stripping (chemical process). Nitrification process is done by bacteria in an additional aeration tank where they convert ammonia to a nitrate that is nontoxic to fish, while denitrification is done by another bacteria (anaerobic) that convert nitrate to nitrogen gas ($N_2$). Ammonia stripping process is done in a stripping tower by adding lime (CaO).

*Phosphorus removal:* Municipal wastewater treatment removes about 30% of the phosphorus by biological processes (Drinan and Whiting, 2001); this treatment is not satisfactory to the receiving body or for reusing water purposes. An additional treatment process, such as chemical precipitation using ferric chloride, lime or alum, can do the purification of water from phosphorus; this process is known as coagulation and sedimentation, where small particles clump together to form large mass before settling.

*Disinfection*: The final step of municipal wastewater treatment is the disinfection to kill any pathogens microbes to protect water receiving body, thus, human and animal health. For water and wastewater disinfection purposes, several processes can be used such as chlorination (using chlorine, chlorine dioxide, bromine chloride or potassium permanganate), ozonation, ultraviolet (UV) radiation systems, air stripping, membrane processes and activated carbon adsorption.

The disinfection by chlorine is the most common one in wastewater treatment because it has a satisfactory efficiency of disinfection and low cost. However, high chlorine concentrations in water (dechlorination) must be reduced before discharging to the environment. The aim of dechlorination is to protect aquatic life from high chlorine concentrations. The dechlorination process uses several chemicals such as sulfur dioxide, sodium sulfite or sodium metabisulfite (Drinan and Whiting, 2001); these can be applied in the form of a gas, liquid or as a solid.

Disinfecting by ultraviolet radiation is more effective than using chlorine. For high quality water, ozone is an excellent disinfectant but it is expensive. However, the efficiency of disinfection treatment by ultraviolet radiation (UV) is reduced by turbidity.

*Regulation of pH*: Before water discharging, effluents must be neutralized to have a pH of seven by adding some chemicals (acids or basics). Next, the effluents can be discharged or backed to the environment via various methods: directly in rivers and seas or indirectly through evaporation or reusing as irrigation.

### 10.4.2.5 Wastewater Residues Management

The second line of wastewater treatment is the treatment of water solid residues. Water solid residues include all types of solid separated from different steps of wastewater treatment. The large solids, such as large floating debris, plastic bags and bottles, leaves, logs and limbs can be separated by the screening process and disposed of at traditional solid waste (landfills or incineration).

The particulate matter and suspended solids, depending on their density, could be separated by sedimentation (the sludge is a settled solids from reactor bottom) and flotation (scum layer from reactor surface); these can be treated as activated sludge. The sedimentation (sludge or activated sludge) from primary and secondary treatment are treated by anaerobic digestion then dewatered, thickened, dried and disposed of at organic solid waste (land application, composting, landfill or incineration). Some authors classify the sludge to two types: coagulation sludge and softening sludge (Drinan and Whiting, 2001).

*The coagulation sludge (known as alum sludge)* is treated by coagulation–sedimentation process. This is usually done by adding specific chemicals such as alum (aluminum sulfate, $Al_2(SO_4)_3$, aluminum chloride, $AlCl_3$, aluminum chlorohydrate, general formula $Al_nCl(3n-m)(OH)_m$), lime or iron salts. This type of sludge is a mixture of organic and inorganic solids and it is considered a chemical sludge (industrial waste). It is not easy to dewater, and needs special treatment for recovery (coagulant recovery) or disposal. It is also considered a hazardous waste that should be disposed in landfills.

*The softening sludge (known as lime sludge)* is produced by precipitation. It is a mixture of organic and inorganic solids. It contains calcium carbonate and magnesium hydroxide and it is easy to dewater (Drinan and Whiting, 2001).

*Sludge treatment*: The activated sludge or bio-solid is produced during the wastewater treatment processes which will be collected then treated. The principal unit of treatment is anaerobic digestion (AD) (see Chapter 7 – biological processes) where energy (biogas) can be recovered. The AD treatment is a stabilization method. The stabilization includes odor control and reducing pathogenic caused by aerobic digestion (composting), chemical and thermal methods. It also includes techniques such as thickening, dewatering, drying and treated with chemicals.

Thickening is a physical process where the solids are concentrated, e.g. by gravity, followed by flotation or centrifugation, and reduce their volume by removing water (dewatering). Dewatering can be done by vacuum or filtration or drying systems, these methods can be applied mechanically, thermally or naturally (open evaporation lagoons or basins).

Heat treatment and the addition of chemicals as lime or other alkaline materials, ferric chloride, ferrous sulfate and polymers can be used in the stabilization of sludge. The chemical addition and heat treatment are known as conditioning processes (Drinan and Whiting, 2001). The solid cake which is formed after these operations can be easily transported and then disposed of in ultimate disposal with various methods such as incineration, landfilling or reused after treatment as compost (see Chapter 8). The application as land fertilizer (land reclamation) or disposal on land requires special treatment and authorization. Currently, water application (ocean dumping) is prohibited by environmental regulations.

## 10.5 Mass Balance (Solid–Liquid Separation)

Figure 10.7 illustrates the mass balance for the activated sludge process in secondary treatment. This process is used in engineering for biological treatment

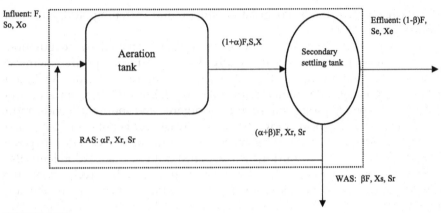

**FIGURE 10.7**

Material balance for solid–liquid separation system (aeration tank).

reactor design, in case of continuous-flow stirred-tank reactor (CFSTR). The process kinetic model is based on the Monod equation and mass balances (biomass, X and substrate, S) in the aeration tank and secondary clarifier. The system boundary is shown by the dashed line in Figure 10.7.

In the aeration tank, the variation in time of the concentration of microorganisms (X) that grow at a rate $\mu_g$ is given by:

$$dX/dt = \mu_g X$$

While the substrate (S) or BOD decays with rate $k_d$:

$$dS/dt = -k_d S$$

where:

    $dX/dt$ = variation of net growth rate of microorganisms, mg MLVSS/l.day

    $dS/dt$ = variation of the substrate (S), mg/l.day

The Monod equation with death rate is:

$$\mu_{net} = \mu_{max} S/(K_s + S) - k_d \qquad (10.1)$$

where:

    $\mu_{net}$ = net growth rate of microorganisms, mg MLVSS/l.day;

    $k_d$ = decay rate of microorganisms, $d^{-1}$;

    $K_s$ = half-velocity constant, mg/l (soluble $BOD_5$ concentration at one-half the maximum growth rate, see biological processes, Chapter 7);

$\mu_{max}$ = maximum growth rate, d$^{-1}$;

S = substrate biodegradable in aeration tank and effluent, mg/l.

Assuming the steady state material balances for biomass, the rate limiting substrates in an activated sludge tank, short residence times in the settling tank and tank is completely mixed. In the aeration tank, the two balances are:
Biomas, X:

$$[\mu_{max} S/(K_s + S) - k_d]XV + \alpha FX_r = (1 + \alpha)FX \qquad (10.2)$$

Substrate, S:

$$FS_0 + \alpha FS_r = 1/Y_{x/s} [\mu_{max} S/(K_s + S)]XV + (1 + \alpha)FX \qquad (10.3)$$

Where:

F = flow rate of wastewater, l.day$^{-1}$

$X_o$ = the concentration of biomass or microorganism (mixed-liquor volatile suspended solids, MLVSS) in the influent, mg/l;

$X_e$ = the concentration of biomass that does not settle in the secondary clarifier, mg/l;

$X_r$ = biomass concentration in the recycled sludge flow, mg/l;

X = biomass concentration in the aeration tank, mg/l;

V = aeration tank volume, l;

$Y_{x/s}$ = The yield coefficient of biomass – substrate.

$\beta F$ = sludge waste rate, l.day$^{-1}$;

$(1 - \beta)F$ = effluent rate, l.day$^{-1}$.

Assuming no separation in the setting tank ($S = S_e = S_r$):

$$\mu_{net} VX = [\beta FX_r + (1 - \beta)FX_e]$$

Solid (cells) residence time:

$$\theta_c = 1/\mu_{net} = VX/[\beta FX_r + (1 - \beta)FX_e]$$

Liquid (hydraulic) residence time:

$$\theta_H = V/F = [\beta X_r + (1 - \beta)X_e]/X\mu_{net}$$

Volume of sludge tank for a certain degree of BOD removal ($S_o - S$):

$$V = Y_{x/s} F (So - S)/ \mu gX = Y_{x/s} \theta_c F (S_o - S)/X(1 + k_d \theta_c)$$

In terms of recycle:

$$V = \theta_c F (1 + \alpha - \alpha X_r / X)$$

## EXAMPLE

A schematic diagram of an activated sludge unit is provided in Figure 10.7. An industrial waste with an inlet $BOD_5$ value of a wastewater feed stream to an activated sludge unit is $S_o = 300$ mg/l must be treated to reduce the exit $BOD_5$ level to be $S = 30$ mg/l. The inlet flow rate is 20,000 m³/h. The kinetic parameters have been estimated for waste as $\mu_{max} = 1.50$ day⁻¹, $K_s = 400$ mg/l of $BOD_5$, $Y_{x/s} = 0.5$ gdw/mg BOD, and $k_d = 0.07$ day⁻¹. Assuming a recycle ratio of $\alpha = 0.50$ and a steady state biomass concentration of $X = 5$ g/l, calculate:

1) Solids residence time ($\theta_c$);
2) Required reactor volume (V);
3) Hydraulic residence time ($\theta_H$);
4) Biomass concentration in recycle ($X_r$);
5) Determine the daily oxygen requirement.

## SOLUTION:

**Data:1**

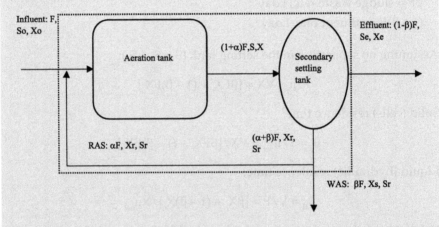

Influent: F, So, Xo

Aeration tank

(1+α)F,S,X

Secondary settling tank

Effluent: (1-β)F, Se, Xe

RAS: αF, Xr, Sr

(α+β)F, Xr, Sr

WAS: βF, Xs, Sr

$$S_o = 300 \text{ mg/l}; S = 30 \text{ mg/l}$$

$$F = 20,000 \text{ m}^3/\text{h}$$

$$\mu_{max} = 1.50 \text{ day}^{-1}; \ k_d = 0.07 \text{ day}^{-1}; \ K_s = 400 \text{ mg/l of BOD}_5$$

$$Y_{x/s} = 0.5 \text{ gdw/mg BOD}; \ \alpha = 0.5; \ X = 5 \text{ g/l}.$$

1) Monod equation can be used with death rate term to estimate the solids residence time $\theta_c$:

$$\theta_c = 1/\mu_{net} \Rightarrow \text{Monod equation is } \mu_{net} = [\mu_{max}S/(Ks + S)] - k_d$$
$$\Rightarrow \theta_c = 1/[[\mu_{max}S/(K_s + S)] - k_d] = 1/[[(1.50 \text{ day}^{-1} \times 30 \text{ mg/l}) / (400 \text{ mg/l} + 30 \text{ mg/l})] - 0.07 \text{ day}^{-1}] = 1/[(45/430) - 0.7] = 1/ 0.03465 = 28.86 \approx 29 \text{ days}$$

2) Equation of reactor volume can be used:

$$V = [Y_{x/s} \ \theta_c \ F \ (S_0 - S)]/X(1 + k_d\theta_c) = [0.5 \text{ gdw/mg} \times 29 \text{ day} \times 24 \text{ h/day} \times 20,000 \text{ m}^3/\text{h} \ (300 - 30) \text{ mg/l}] / [5 \text{ g/l} \times (1 + 0.07 \text{ day}^{-1} \times 29 \text{ day})] = 1.2404 \times 108 \text{ m}^3$$

3) Hydraulic residence time $(\theta_H)$:

$$\theta_H = V/F = 1.2404 \times 108 \text{ m}^3 / 20,000 \text{ m}^3/\text{h} = 6202 \text{ h} \approx 258 \text{ days}$$

4) Biomass concentration in recycle $(X_r)$:

We could use the equation reactor volume in terms of recycle ratio

$$V = F\theta_c[1 + \alpha - \alpha(X_r/X)] \Rightarrow X_r = [(V/F\theta_c) - (1 + \alpha)]X/\alpha =$$

$$[[1.2404 \times 108 \text{ m}^3/(20,000 \text{ m}^3/\text{h} \ 29 \text{ day} \times 24 \text{ h/day})] - (1 + 0.5)] \times (5 \text{ g/l}) / (0.5) = 74.1 \text{ g/l}$$

5) Determining the daily oxygen requirement:

To get the oxygen demand (over five days) in milligrams:

The $BOD_5$ removal daily $= F \ (S_0 - S) = 20,000 \text{ m}^3/\text{h} \ (0.300 - 0.030)$ g/l $\times 1,000 \text{ l/m}^3 \times 1/24 \text{ h/day} = 2,25,000 \text{ g/day} = 225 \text{ kg/day}$

## Review Questions

Briefly describe the contaminants of wastewater.
What are the differences between black water and gray water?
Differentiate between sanitary wastewater and non-sanitary wastewater.
Differentiate between on-site treatment and off-site treatments.
Mention the main steps of wastewater treatment.
Explain the management system of industrial wastewater.
Mention the wastewater treatment processes.
Describe the municipal wastewater treatment processes.

## References

1. S. Al Arni, 2014, "Treatment and Recycling of Water Resulting from the Ablution and Homes for Help in Solution of the Water Crisis in Most Countries of the Islamic World" Journal of King Saud University: Engineering Science, Vol. 26, pp. 15–36.
2. M. Nadjafi, A. Reyhani and S. Al Arni, 2018, "Feasibility of Treatment of Refinery Wastewater by a Pilot Scale Hybrid Membrane System (MF/UF&UF/RO) for Reuse at Boilers and Cooling Towers" accepted for publication in the Journal of Water Chemistry and Technology, Vol. 40, Supplement 3, pp. 167–176.
3. S. Al Arni, J. Amous and D. Ghernaout, 2019, "On the Perspective of Applying of a New Method for Wastewater Treatment Technology: Modification of the Third Traditional Stage with Two Units, One by Cultivating Microalgae and Another by Solar Vaporization", International Journal of Environmental Sciences & Natural Resources (IJESNR), Vol. 16, Supplement 2, available online at https://juniperpublishers.com/ijesnr/IJESNR.MS.ID.555934.php
4. A. Converti, A. Del Borghi, S. Arni and F. Molinari, 1999, "Linearized Kinetic Models for The Simulation of the Mesophilic Anaerobic Digestion of Pre-Hydrolyzed Woody Wastes", Chemical Engineering & Technology, Vol. 22, Supplement 5, pp. 429–437.
5. V.K. Gupta and I. Ali, 2013, Environmental Water Advances in Treatment, Remediation and Recycling, Elsevier, ISBN: 978-0-444-59399-3.
6. A. Converti, G.L. Mariottini, A.M. Ben Hamissa, E. Finocchio, S. Al-Arni, R. Botter and A. Lodi, 2013, "Cadmium Removal from Aqueous Solutions by Biosorbents: Study of the Operating Conditions" in M. Hasanuzzaman and M. Fujita (eds.) "Characteristics, Sources of Exposure, Health and Environmental Effects", Series: Chemistry Research and Applications, Nova Science Publishers, Inc., Hauppauge, NY, USA—3rd Quarter, ISBN: 978-1-62808-722-2 (Chapter 9: Pages 213–233).
7. G. Boari, I.M. Mancini and E. Trull, 1997, "Technologies for Water and Wastewater Treatment", CIHEAM – Options Méditerranéennes, Sér. Séminaires Méditerranéens, Vol. 37, pp. 261–287.

8. A.P. Sincero and G.A. Sincero, 2003, "Physical–Chemical Treatment of Water and Wastewater", CRC Press, ISBN: 1-58716-124-9.

9. G. Tchobanoglous, F.L. Burton and S.H. David, 2003, "Wastewater Engineering: Treatment and Reuse", Metcalf and Eddy, Inc., 4th Edition, McGraw-Hill, EISBN: 0-07-112250-8.

10. A. Converti, A. Del Borghi, M. Zilli, S. Arni and M. Del Borghi, 1999, "Anaerobic Digestion of the Vegetable Fraction of Municipal Refuses: Mesophilic versus Thermophilic Conditions", Bioprocess Engineering, Vol. 21, pp. 371–376.

11. A.R. Mels and E. Teerikangas, 2002, "Physico–Chemical Wastewater Treatment" in P. Lens, L.H. Pol, P. Wilderer and T. Asano (eds.) "Water Recycling and Resource Recovery in Industry: Analysis, Technologies and Implementation", IWA Publishing, Alliance House, London, ISBN: 1 84339 005 1.

12. Tchobanoglous, Burton and Stensel, "Wastewater Engineering: Treatment and Reuse".

13. T. Asano, F.L. Burton, H.L. Leverenz, R. Tsuchihashi and G. Tchobanoglous, 2007, "Water Reuse: Issues, Technologies, and Applications", Metcalf & Eddy (AECOM), McGraw-Hill, New York, ISBN-13:978-0-07-145927-3.

14. G.M. Kemp, 2008, "Pretreatment Program Requirements for Industrial Wastewater" in M.D. Nelson (ed.) "Water Environment Federation (WEF), Operation of Municipal Wastewater Treatment Plants, Manual of Practice No. 11", 6th Edition, Vol I, Management and Support Systems, McGraw-Hill, New York. DOI: 10.1036/0071543678.

15. A.D. Levine and T. Asano, 2002, "Water Reclamation, Recycling and Reuse in Industry" in P. Lens, L.H. Pol, P. Wilderer and T. Asano (eds.) "Water Recycling and Resource Recovery in Industry: Analysis, Technologies and Implementation", IWA Publishing, Alliance House, London, ISBN: 1 84339 005 1.

16. J. Lipták and D.H.F. Liu, 1999, "Screens and Comminutors", subchapter (Chapter7 – Wastewater Treatment) in D.H.F. Liu and B.G. Liptak (eds.), "Environmental Engineers' Handbook", CRC Press LLC, Boca Raton, FL, ISBN-10:0849321573.

17. J.E. Drinan and N.E. Whiting, 2001, "Water and Wastewater Treatment: A Guide for the Nonengineering Professional", CRC Press LLC, Boca Raton, FL, ISBN: 1-587 16-049-8.

18. W.K. Shieh and V.T. Nguyen, 1999, "Disinfection", subchapter (Chapter7 – Wastewater Treatment) in D.H.F. Liu, B.G. Liptak (editors), "Environmental Engineers' Handbook", CRC Press LLC, Boca Raton, FL, ISBN-10:0849321573.

19. L.H. Reuter, 1999, "Conventional Sewage Treatment Plants", subchapter (Chapter7 – Wastewater Treatment) in D.H.F. Liu and B.G. Liptak (eds), "Environmental Engineers' Handbook", CRC Press LLC, Boca Raton, FL, ISBN-10:0849321573.

20. F.P. Sebastian, 1999, "Advanced or Tertiary Treatment", subchapter (Chapter7— Wastewater Treatment) in D.H.F. Liu and B.G. Liptak (eds.), "Environmental Engineers' Handbook", CRC Press LLC, Boca Raton, FL, ISBN-10:0849321573.

# 11

## Effluent Gas Control: Clean-up System

### Key Learning Objective

- Understanding the effluent gas treatment process.

### 11.1 Introduction

Emission of pollutants to the atmosphere is established and regulated by environmental legislations that set emission limits. These limits are very similar in most countries of the world and they are independent of the operation conditions of the plant, but they must be met by the efficiency of treatment. For instance, in KSA, Saudi Arabian Presidency of Metrology and Environment (PME) established an air pollution guidelines and standards [1]. Table 11.1 shows the standard of ambient air quality in Saudi Arabia as presented by PME [2].

There are different industrial sources of pollutant emissions such as industrial incinerators (stacks). The typical volume of flue gas produced by MSW incinerators is between 4,000 and 6,000 $m^3$ per ton of waste [3,4]. The major types of pollutants emitted by combustion plants take the form of particulates and gaseous; these contain carbon oxides ($CO_x$), nitrogen oxides ($NO_x$), water vapor ($H_2O$), sulfur oxides ($SO_x$), hydrogen chloride (HCl), hydrogen fluoride (HF) and other hydrocarbons such as polycyclic aromatic hydrocarbons, dioxins and furans and heavy metals such as mercury, cadmium and lead. Table 11.2 shows the standard of stack gases related to boilers and industrial,

**TABLE 11.1**

Air quality standards outdoor of a pollution source for industrial areas

| Pollutant | Maximum concentration at 25 °C and 760 mmHg, µg/Nm³ (ppm) |
|---|---|
| Ammonia (NH₃) | 1,800 (2.6) hourly average |
| Benzene (C₆H₆) | 5.0 (0.0015) annually average |
| Carbon Monoxide (CO) | 40,000 (32) hourly average for maximum twice a month |
| | 10,000 (8.1) 8-hour average for maximum twice a month |
| Chlorine (Cl₂) | 300 (0.1) hourly average |
| Fluorides | 1.0 (0.001) monthly average |
| Hydrogen Sulfide (H₂S) | 150 (0.1) hourly average for maximum once per year |
| | 40 (0.03) daily average for maximum once per year |
| Inhalable particulates (PM₂.₅) | 35 daily average (The average 90th Percentile 24-hour |
| (less than 2.5 microns | concentration must not exceed 35 µg/Nm³) |
| equivalent aerodynamic | 15 annually average |
| diameter) | |
| Inhalable Particulates (PM₁₀) | 340 daily average (The average 90th Percentile 24-hour |
| (less than 10 microns | concentration must not exceed 340 µg/Nm³) |
| equivalent aerodynamic | 80 annually average |
| diameter) | |
| Lead | 1.5 three-months average |
| | 0.5 (0.00005) annually average |
| Nitrogen dioxides (NO₂) | 660 (0.35) hourly average for maximum twice a month |
| | 100 (0.05) annual average |
| Non-methane Hydrocarbon | 160 (0.24) 3-hour average |
| (NMHC) | **Note:** There is no adopted standard for NMHC; this level is a |
| | goal to aid in the control of ambient ozone concentrations. |
| | Sampling period 0600–0900 hours |
| Ozone (O₃) | 235 (0.12) hourly average for maximum 2 times a month |
| | 157 (0.08) 8-hour average for maximum 2 times per 7 days |
| Sulfate | 25 daily average |
| Sulfur Dioxide (SO₂) | 730 (0.28) hourly average for maximum 2 times per year |
| | 365 (0.14) daily average for maximum once time per year |
| | 80 (0.03) annually average |

Source: Royal Commission for Jubail and Yanbu, 2004, "Royal Commission Environmental Regulations (RCER)" Vol I, PME, 2012, "Standards and Guidelines for air pollution".

furnaces burning and hazardous materials. Table 11.3 shows the standards of incinerator related to hazardous and medical waste incineration.

Industrial control of the discharges of atmospheric pollutants could be arranged before and after production process by minimizing emissions of pollutants; this makes an integral part of the design of the plant and considered, after emissions, as a treatment technology and efficient removal.

This chapter is dedicated to the treatment technologies of pollutants from gas streams after production and before discharge to the air. Gaseous waste from fuel cycle facilities, reactors vitrification, incineration, pyrolysis and gasification can be cleaned by filtration, adsorption and absorption. A gas

**TABLE 11.2**

Standards of stack gases (boilers and industrial furnaces burning hazardous materials)

| Pollutant | Emission limits* | |
|---|---|---|
| Organic Emissions | 99.99% destruction removal efficiency | |
| CO | Not to exceed 100 ppmv on an hourly rolling average basis, corrected to 7% oxygen, dry gas basis | |
| $NO_x$ | As combustion device standards (new and modified facilities not combusting chlorinated organics) | |
| $SO_2$ | As combustion device standards (new and modified facilities) | |
| Particulate | 180 mg/dscm[1] after correction to 7% oxygen stack gas concentration | |
| Chlorinated organics | 99.9999% destruction removal efficiency | |
| Metals | Ag | $1.5 \times 104$ g/h |
| | As | 11 g/h |
| | Ba | $2.5 \times 105$ g/h |
| | Be | 21 g/h |
| | Cd | 28 g/h |
| | Cr | 4.2 g/h |
| | Hg | 1,500 g/h |
| | Pb | 430 g/h |
| | Sb | 1,500 g/h |
| | Tl | 1,500 g/h |

\* Compliance with the standards will be determined by comparison with hourly average data, unless otherwise specified, that are corrected to standard temperature and pressure, moisture and oxygen content as specified by USEPA Methods.
(1) Milligrams per Dry Standard Cubic Meter

Source: Royal Commission for Jubail and Yanbu, 2004, "Royal Commission Environmental Regulations (RCER)" Volume I.

clean-up system depends on the type and quantity of waste in the gas effluent and the emission limits to atmosphere and the treatment efficiency required.

## 11.2 Gas Clean-up Systems

Treatment technologies of gaseous waste classify the pollutant emissions on the basis of source temperature into low and high temperature emissions:

- Sources at *low temperature* such as production processes and work-up of products (handling and storage activities that cause emissions). The emissions may include solid raw materials, volatile organic compounds (VOCs), inorganic volatile compounds or both, with or without dusty content.

**TABLE 11.3**

Standards of incinerator (hazardous and medical waste incineration)

| Pollutant | Emission limits* |
|---|---|
| Particulate | 34 mg/dscm corrected to 7% oxygen |
| Visible emissions | 10% opacity except for no more than 6 minutes in any hour |
| Sulfur dioxide | 500 mg/dscm |
| CO | 100 mg/dscm (existing facilities) |
|  | 50 mg/dscm (new and modified facilities) |
| Organics | >99.99% destruction removal efficiency (DRE) for each organic constituent (existing facilities) >99.9999% (new and modified facilities) |
| Total dioxins and furans | 30 ngTEQ/dscm @ 7% oxygen (existing facilities) |
|  | 0.1 ngTEQ/dscm (new and modified facilities) |
| PCB | 1 mg/kg PCB feed for a maximum one-hour average concentration or >99.9999% destruction removal efficiency (DRE) |
| Hydrogen chloride | 100 mg/dscm OR at least 99% removal efficiency if emission is > 1.8 kg/h (existing facilities) |
|  | 10 mg/dscm (new and modified facilities) |
| Hydrogen fluoride | 5 mg/dscm (existing facilities) |
|  | 1 mg/dscm (new and modified facilities) |
| Metals | Ag      3000 g/h (existing facilities) |
|  | As      2.3 g/h (existing facilities) |
|  | Ba      0.5 mg/dscm (new and modified facilities) |
|  | Be      50,000 g/h (existing facilities) |
|  | Cd      4.0 g/h (existing facilities) |
|  | Co      5.4 g/h; (existing facilities) |
|  | Cr      0.05 mg/dscm (new and modified facilities) |
|  | Cu      0.5 mg/dscm (new and modified facilities) |
|  | Hg      0.82 g/h; (existing facilities) |
|  | Mn      0.5 mg/dscm (new and modified facilities) |
|  | Ni      0.5 mg/dscm (new and modified facilities) |
|  | Pb      300 g/h (existing facilities) |
|  | Sb      0.05 mg/dscm (new and modified facilities) |
|  | Tl      0.5 mg/dscm (new and modified facilities) |
|  | V      0.5 mg/dscm (new and modified facilities) |
|  | 90 g/h (existing facilities) |
|  | 0.5 mg/dscm (new and modified facilities) |
|  | 300 g/h (existing facilities) |
|  | 0.5 mg/dscm (new and modified facilities) |
|  | 300 g/h (existing facilities) |
|  | 0.05 mg/dscm (new and modified facilities) |
|  | 0.5 mg/dscm (new and modified facilities) |
| Incineration chamber: minimum post combustion temperature and minimum residence time | 850 °C for 2 seconds (1 second for existing facilities) OR 1100 °C for 2 seconds where incineration of >1% halogenated organic substances (expressed as chlorine) takes place |

\* Compliance with the standards will be determined by comparison with hourly average data, unless otherwise specified that are corrected to standard temperature and pressure, moisture and oxygen content as specified by USEPA Methods

TEQ = toxic equivalent

Source: Royal Commission for Jubail and Yanbu, 2004, "Royal Commission Environmental Regulations (RCER)" Volume I.

- Sources at *high temperature* such as combustion processes and their facilities (power plants, incinerators, boilers, catalytic oxidizers and thermal). The emissions may include a mixture of particulate matter or dust, heavy metals, carbon oxides, nitrogen oxides, sulfur oxides, halogen compounds such as HCl, HF and $Cl_2$, dioxins and furans. For example, air pollutants of incineration include gases, solid particulate or aerosol and their associated forms including dusts, vapors, fogs, mists, fumes and smokes.

There are several methods of cleaning gaseous pollutants from gas streams; examples include absorption, adsorption, condensation, incineration and bio-filtration [5].

The aim of using the treatment techniques is to reduce the emission of air pollutants to atmosphere. The treatment technologies depend on the types of pollutants in the waste stream. Air pollutants have different physical states and chemical compositions and they originate from different sources.

## CALCULATION OF FLUE GAS VOLUME

Kristalina and Keshav (1999) presented a method for calculating the flue gas volume. The calculation of flue gas volume depends on temperature, pressure and composition of the gas. On the basis of standard or normal condition of T & P (T = 0 °C, P = atmosphere), the flue gas volume can be calculated. As a rule of thumb, it is assumed that the flue gas is dry, has zero oxygen content and stoichiometric flue gas quantity is 0.25 $Nm^3$ per MJ. The equation for gas volume calculation is:

$$2H_{inf}[(273 + t)]/[(21 - y)(11 - z)] \ [m^3/kg]$$

Where:

$H_{inf}$ = Calorific value of fuel [MJ/kg];
t = temperature [°C];
y = percent of $O_2$;
z = percent of $H_2O$

The development of the equation is:

Stoichiometric flue gas volume = $H_{inf} \times 0.25 \ [Nm^3/kg]$ (dry, 0% $O_2$)

Dry flue gas at y% $O_2$ = $H_{inf} \times 0.25 \times (21/(21 - y)) \ [Nm^3/kg]$

Wet flue gas at z% $H_2O$ = $H_{inf} \times 5.25 /(21 - y) \times (100/(100 - z)) \ [Nm^3/kg]$

Actual flue gas volume at t °C = $H_{inf} \times 525 /(21-y) / (100-z) \times$
$(273+t)/273 =$

$$2H_{inf}[(273+t)]/[(21-y)(11-z)] \ [m^3/kg]$$

**Example:** Calorific value $H_{inf}$ = 8 MJ/kg, oxygen content y = 11%, water vapor content z = 15%, flue gas temperature t = 100 °C.

Actual flue gas volume: $2 \times 8 [(273+100)]/[(21-11)$
$(100-15)] = 7.0 \ [m^3/kg]$

**Note:** For more information about calculations, a recommended book is: C.C. Lee and Shun Dar Lin (editors), 2007, "Handbook of Environmental Engineering Calculations", 2nd edition, McGraw-Hill, ISBN: 0071475834.

The abatement techniques used for the treatment of waste gases include, as examples, flaring, thermal oxidation, catalytic oxidation, gravity separators, cyclones, electrostatic precipitators, mist filters (such as 2-stage dust filters, fabric filters, catalytic filtration, absolute filters, high efficiency air filter), dry, semi-dry and wet sorption, bio-filtration, bio-scrubbing, bio-trickling, selective non-catalytic reduction (SNCR) and selective catalytic reduction (SCR).

On the other hand, recovery techniques also include gravity separators, cyclones, membrane separation, condensation, adsorption, wet scrubbers, dry, semi-dry and wet sorption, mist filters (as 2-stage dust filters, fabric filters, absolute filters, high efficiency air filter), electrostatic precipitator (ESP).

The treatment techniques, used in the cases of low-temperature sources, firstly remove solid material or mists, then remove the gaseous pollutants. If necessary, further abatement can be made that is applied on the basis of emission levels required.

The treatment techniques depend on the separation technology, the method of operation, the operating conditions and the efficiency of treatment required. The separation technology depends on the properties of gas stream as chemical–physical characteristics, which include type of components, temperature, pressure, density, viscosity, surface tension, electrostatic forces, particle size and several other properties.

## 11.3  Types of Waste Gases and Treatment Methods

In the following, a brief description of various types of waste gases and their control and treatment methods will be presented.

*Dusts*: Dusts are particulate materials found in the gas's effluents. They could be organic or inorganic (metallic or non-metallic as silica) and aerosols/droplets. The particulates in the flue gases usually have size range between 1 and 75 µm. The particulates control of flue gases depend on the particle load in the gas stream, the flowrate of gas, the average particle size, the particle-size distribution, the flue gas temperature and the required outlet gas concentration. Based on these parameters, the selection of gas cleaning equipment can be determined.

Dust can be removed from waste gas streams by adequate techniques according to the actual situation. In certain cases, pretreatment processes become necessary in order to protect equipment that are used in successive treatment phases. The pretreatment could include gravity separation which could help in the prevention of filter clogs in successive filtering process.

The combustion process produces a very dusty waste with heterogeneous and toxic nature, usually composed of ash, and requires efficient and high levels of pollutants removal. Also associated with dust particulates, there are more toxic pollutants such as dioxins and furans and heavy metals. Particulate emissions from incinerators can be collected by cyclones, electrostatic precipitators and fabric filters.

*Soot:* Soot is a black powder or flake substance consisting of a large amorphous carbon that is produced by the incomplete burning of organic matter in conditions of high temperature and low oxygen content. Soot contains molecules such as acetylene radicals and polycyclic aromatic hydrocarbons (PAHs). The reduction of soot formation can use secondary combustion air to complete the burning of carbon that is not burned.

*Smuts*: Smut is a small flake of soot or smudge that is a product of a combustion that contains corrosive acids such as hydrochloric, sulfuric or hydrofluoric acid.

*Heavy metals*: Heavy metals (such as arsenic (As), cadmium (Cd), chromium (Cr), cobalt (Co), copper (Cu), mercury (Hg), manganese (Mn), nickel (Ni), lead (Pb), antimony (Sb), thallium (Tl), vanadium (V), zinc (Zn), etc.) are produced from incinerators as flue gas, fly ash and bottom ash during the combustion of waste process. They are released during the combustion process as ash metals particulate and/or metal compounds or their derived compounds, and they are a function of their physio-chemical properties (e.g. the effect of high temperature, depending on their volatility). Heavy metals may also react with oxygen or hydrogen chloride, and/or other compounds or elements to form other compounds or as adsorption onto the fine particulates. During the processes after combustion of waste, as the temperature of flow gas decreases, the metals volatilization will condensate. Then, they will be associated with the particulate and with their abatement.

The recycling of batteries reduces the emissions of cadmium and mercury. Because heavy metals have a toxic health effects, they need to be removed from waste and waste gases; electrostatic precipitators and fabric filters can be used for abatement. For detailed information about heavy metals and their effects on human health, see Chapter 2.

*Volatile organic compounds (VOCs)*: Treatment technologies of VOCs depend on the types of pollutants in the waste stream and the final aim of treatment which is recovery or destruction. A choice of a technique or another depends on the pollutant characteristics influencing the emission stream. In the case of destruction, flares can be used. Flares are simply a vent to atmosphere where the flammable gases can be burned; the process of destruction is a thermal oxidation with efficiency of over 99%, used, usually, for continuous emission streams in oil refineries, power stations and other several industries. Sometime before using this system, another recovery technique or preliminary treatment can be used; these include cryogenic systems at low temperature or refrigeration and condensation techniques.

Destruction process is also used in case of maintenance or upset systems. It usually uses flaring to dispose gases safely without connecting the surplus combustible gases to abatement systems. Furthermore, biological treatment can be applied to remove VOC from waste gas streams. If the recovery of VOCs has a commercial value, other techniques can be used; these include condensation, membrane separation or adsorption by wet scrubbing on the solid material then regenerated. The condensation techniques are usually used as an integral part of tanks for pollution control. Adsorption by activated carbon is the effective technology for removal of organic pollutants and VOCs at low temperatures. The microporous aluminosilicate minerals (Zeolites) are also used.

*Acidic and corrosive pollutants*: Municipal waste incinerators produce acidic and toxic or corrosive gases such as hydrogen chloride, hydrogen fluoride and sulfur dioxide. Chlorine in flow gas can be derived from plastic material waste such as PVC (polyvinylchloride) or other source materials such as rubber, leather, paper and vegetable matter that contain metal chlorides like NaCl or $CaCl_2$. On the other hand, fluorine can be derived from PTFE (polytetrafluoroethylene) waste materials, while sulfur dioxide can be derived from combustion of sulfurous compounds. The nitrogen oxides ($NO_x$) can be derived from part of the nitrogen in the air and the nitrogen in the waste.

For the removal or recovery of acidic pollutants such as hydrogen chloride (HCl), hydrogen fluoride (HF), sulfur dioxide ($SO_2$) and nitrogen oxides ($NO_x$) from waste, gas stream can use scrubbers in three different techniques; these are the wet, wet-dry and dry-scrubbing.

*Nitrogen oxides (NOx)*: Nitrogen oxides ($NO_x$) emissions in flue gases consist of two types: the first type is produced by waste incineration (known as fuel $NO_x$) and the second is produced by oxidation of nitrogen present in air at high temperatures (known as thermal $NO_x$). The thermal $NO_x$ represents about 25% of the total nitrogen oxides [6].

In large combustion plants, these are typically composed of about 90% NO and 10% $NO_2$ (Williams, 2005). Furthermore, wastewater treatment and agricultural industries are additional sources of nitrogen oxides ($NO_x$) emissions. Nitrogen oxides affect health and environment by the formation of photochemical smog and their contribution to acid rain. These oxides ($NO_x$) can be reduced by recirculation of the flue gases into the combustion chamber or by ammonia addition or by bioprocesses [7].

*Sulfur dioxide ($SO_2$) treatment or Fuel Gas Desulfurization (FGD)*: FGD techniques are mainly used to treat gases with low concentration of sulfur dioxide (up to 1%) which are usually produced by power plants and large combustion plants. Dry or semi-dry scrubber techniques can use lime or limestone to produce gypsum (calcium sulfate, $CaSO_4$) according to the following equation:

$$SO_2 + CaCO_3 => CaSO_4 + CO_2 \qquad (11.1)$$

*Dioxins and furans*: Dioxin and furan are organic compounds that combine with chlorine atoms to formulate a family of organic compounds called polychlorinated dibenzodioxins (PCDD) and polychlorinated dibenzofurans (PCDF) (Figure 11.1). All of these compounds are toxic and have health effects to human, animals and environment (Worrell and Vesilind, 2012, [8]; Williams, 2005). The European Community (EC) Waste Incineration Directive, 2000, determined the emission-limit value for old municipal solid waste incinerator for these compounds PCDD and PCDF at 0.1 ngTEQ/$m^3$, while for the modern plants the emissions limit value is between 0.0002 and 0.08 ngTEQ/$m^3$.

Dioxin molecule          Furan molecule

2,3,7,8-tetrachlorodibenzo-p-dioxin     2,3,7,8-Tetrachlorodibenzofuran

**FIGURE 11.1**

Dioxin and furan molecules structure and their congeners with chlorine atoms.

The main sources of dioxins are the waste and the combustion processes. The CO concentration is an indicator on the formation of dioxins in a combustion operation. The relationship between the temperature and CO concentration is given by the following empirical equation (Worrell and Vesilind, 2012):

$$PCDDs + PCDFs = a - bT + cCO \tag{11.2}$$

Where:

a, b and c are constants

a = 2,670.2 and 4,754.6 for modular combustors and water-wall combustors, respectively;

b = 1.37 and 5.14 for modular combustors and water-wall combustors, respectively;

c = 100.06 and 103.41 for modular combustors and water-wall combustors, respectively.

T = the temperature (in degrees Celsius) in the secondary chamber for modular combustors and the furnace temperature in water-wall combustors, respectively.

CO = the concentration of CO in percent of gases.

The production of PCDDs is proportional to the CO concentration according to the following relationship:

$$PCDDs = (CO/A)^2 \tag{11.3}$$

Where:

PCDDs = concentration of dioxins in the off-gases, $ng/m^3$;

CO = concentration of carbon monoxide in the off gases as percent of total gas;

A = a constant, function of operation system.

The emissions of dioxins and furans can be reduced by thermal post combustion, combustion conditions, reducing of the organic contents and active coke technique. In addition, dioxins can be destroyed at high temperature above 850 °C in the presence of oxygen or by catalytic oxidation systems or the process of de-novo synthesis (biological processes) or adsorbed onto solid matter (as activated carbon) by scrubber or filter.

*Other compounds*: For recovery or treatment purposes, other compounds of waste gas pollutants can be treated by the appropriate techniques, such as by using wet scrubbing by aqueous solvent, acidic or alkaline solution for hydrogen halides such as $Cl_2$, $SO_2$, $H_2S$, $NH_3$; or dry scrubbing to treat $CS_2$, COS; or adsorption for $CS_2$, COS, Hg; using incineration to destroy $H_2S$, $CS_2$, COS, HCN, CO; using SNCR/SCR for NOx; using biological gas treatment for $NH_3$, $H_2S$, $CS_2$.

*Odor problem*: It can be managed by chemical stripping, thermal destruction and bio-filtration. The applied techniques include scrubber, fan or flare.

## 11.4 Main Emissions Control Equipment

The most common devices for collection and removal of particles from waste gases are gravity separators, cyclone separators, electrostatic precipitators, fabric filters and wet scrubbers. Table 11.4 provides a guideline for the selection of appropriate air pollution control separation equipment for the treatment of gas streams. Efficiency of treatment depends on the selected device type and the end goal of treatment. Figure 11.2 shows an illustration of some examples of common devices used in the separation technology and fluid cleaning.

The following is a brief description of the main emission control equipment that is used in the stream of waste gases.

*Gravity separators*: Gravity settling chamber (Figure 11.2) is a simple device that is usually used as a preliminary technique in various filter systems to prevent entrainment. It is normally integrated with other equipment and used for recovery materials and it is not restricted for dust content, particulate size more than $PM_{50}$, but also down to $PM_{10}$. The efficiency of pollutant removal depends on particle size and feed concentration; it usually has a range between 10 and 90%. The application limit of flow rate is up to

**TABLE 11.4**

Equipment selection guideline for gas cleaning

| Particle Size | Solids Particulates | Liquid Particulates | Combination |
|---|---|---|---|
| Submicron: <1 micron | Cartridge filter | Scrubber | Scrubber |
| Invisible: 2–50 microns | Coarse (Grit) filter, cyclone | Cyclone, baghouse | Mist and fume, filters |
| Tiny: 51–850 microns | Gravity settler, cyclone | Gravity settler | Fume filters, scrubber |
| Small: 850 microns | Gravity settler | Cyclone | Fume filters, scrubber |

Source: Modified after Cheremisinoff, 2000 (N.P. Cheremisinoff, 2000, "Handbook of Chemical Processing Equipment", Butterworth-Heinemann, ISBN: 0-7506-7126-2).

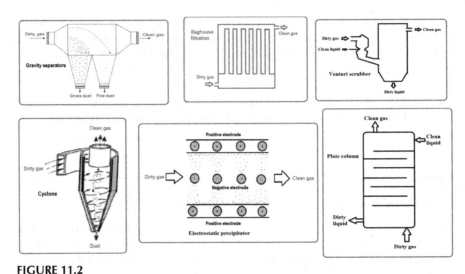

**FIGURE 11.2**

Examples of common devices used in waste particulate emission control.

1,00,000 Nm³/h, but it is effective at low gas velocities of less than 200 m/ min (Hocking, 2005, [9]) in case of particle size in the range of 50–100 μm for a removal efficiency of 90%.

*Cyclone (dry and wet)*: Cyclone (Figure 11.2) is a gravity separator supported by centrifugal forces that the waste gas stream enters tangentially to make a vortex and rotates in a helical path down. The particles are drop down to the bottom and gas flows up through the central inner. The cyclone is usually used after processes of crushing, grinding, spray drying and calcining operations and can be used for recovery. It is often used as pre-cleaners for electrostatic precipitator or fabric filters and suitable for flue gas treatment.

The application limits of temperature can reach more than 1,200 °C, depending on vessel material type. It has an advantage that it can be operated at high temperatures that exceed 500 °C. It can operate at a flow rate of up to 1,00,000 Nm³/h for single unit or up to 1,80,000 Nm³/h for multiple units with a gas concentration up to 16,000 g/Nm³. Typically, the exit gas particulate concentrations of cyclones is in the range of 200–300 mg/m³. The efficiency of pollutant removal depends on the particulate size; it is more effective for removing particles larger than 15 μm, but less effective to finer particles, e.g. PM 80–99%; PM₁₀ 60–95; PM₅ 80–95%; PM₂.₅ 20–70%.

*Electrostatic precipitator* (ESP) (dry and wet): ESP is similar to gravity or centrifugal separator (Figure 11.2), but in ESP the separation treatment is done by electric field (electrostatic force) with high voltage. When the waste gas passes through high voltage electric field, which provides the particles (solid or liquid) with an electrostatic charge, the discharge electrode will have a negative charge while the collecting plate will have a positive charge. The principal cleaning of dust by an electrostatic precipitator has a potential of 50 kV (Nesaratnam and Taherzadeh, 2014) [10].

ESP process consists of three steps: particle charging, particle collection and removal of the collected dust. The ESP can be used to remove particulate matter in industrial applications, chemical manufacturing, refineries, incineration such as utility and industrial power boilers, flue gas streams exiting from waste incineration, cement kilns and glass furnaces, catalytic crackers, paper mills, metal processing, and a wide variety of other processes.

ESP is mainly used to treat small particles in the range from 0.1 to 1 micron [11]. Depending on the type of device, the treatment capacity could reach a very high flow rate of gas volumes between $10^4$ and $4{\times}10^6$ $m^3h^{-1}$ (Nesaratnam and Taherzadeh, 2014). Typically, the exit gas particulate concentrations of ESP are in the range of 5–25 $mg/m^3$. The device can be operated at temperatures up to 700 °C for dry or less than 90 °C for wet matters. For the control of dioxin emissions under lowest possible level, the temperature of operation must be below 200 °C [12]. The efficiency of treatment is typically 80 to 90% and, sometimes, it could be more than 90% for particulate size of 0.1 μm, depending on the type of device. The single-field ESP can reduce the particle concentration to below 150 $mg/Nm^3$, while the two-field ESP is more efficient.

*Scrubber*: Scrubber is a common device used to remove particles or fly ash and toxic gases from flue gases streams. Depending on the properties of waste gas streams and the process required, there are three types of scrubbers: wet, semi-dry and dry scrubber.

Based on the flow rate and operation parameters, there are also three types of scrubber geometries; these are crossflow, counterflow and co-flow scrubbers [13]. The venturi scrubber is an example of co-flow scrubber. In addition, there are many designs of scrubbers such as a fluidized-bed scrubber, a packed scrubber, a tower scrubber and a spray scrubber. A brief description of the main types of scrubbers is provided below.

*1-Wet scrubber*: Wet scrubber is a spray scrubber; it is the most common device used for air pollution control in industries (Figure 11.2). It is used for the removal of toxic gases or particles from waste gases. It can also be used for the treatment of a single target pollutant or as a multipurpose removal. The theory of wet scrubber is based upon the absorption, the mass transfer from gaseous phase into liquid phase (acidic gases in an alkaline liquid phase, as a caustic solution), with or without reaction, which depends on the alkaline solid used such as calcium, magnesium or sodium gypsum slag.

When using two-stage wet scrubbing, in the first stage, water or acidic solution is used as a scrubber medium to remove HF and HCl. After air injection in the second stage, calcium carbonate suspension is used to remove $SO_2$ as calcium sulfate. The HCl and calcium sulfate can be recovered. On the other hand, two-stage wet scrubbing is used without material recovery to separate HCl and HF ions before desulfurization.

The efficiency of wet scrubber process for acid gas removal is very high, for example, it is more than 95% for hydrogen chloride. For abatement of heavy metals as lead, the efficiency is about 99% and for cadmium it is about 92% (Williams, 2005).

*2-Semi-dry scrubber:* The semi-dry scrubber uses a spray absorption process where the droplets of calcium hydroxide cool and neutralize the hot flue gas. When the waste gas passes through the scrubber tower, the reaction takes place between calcium hydroxide and acid gases. It is used in conjunction with fabric filter. It can also be used to treat heavy metals as mercury and cadmium and organic micropollutants as dioxins when activated carbon is added to the calcium hydroxide and furans.

*3-Dry scrubbers:* The dry scrubber method is an absorption process for the treatment of acid gases that pass through a fine-grained alkaline powder as dry calcium hydroxide ($Ca(OH)_2$). It is sprayed onto the gases. The reaction takes place at about 160 °C in the dry state and the products (e.g. calcium chloride, calcium sulfate with hydrogen chloride and sulfur dioxide) are removed from the flue gas stream by a filter. Like the semi-dry scrubber, it could be also used to treat heavy metals.

*Fabric filters:* Fabric filter or baghouses collectors (Figure 11.2) is a device used to separate particulate and acid gases (such as $SO_2$ and HCl) from gas stream, usually used after the scrubbers or the electrostatic precipitator. It is also used to remove organic micropollutants such as dioxins and furans, and heavy metals such as mercury. The mechanism of operation is as follows: the waste gas is introduced through a porous layer of fabric (medium) or fabric bags and it retains the particulates on the medium surface. After collection, the dust cake is removed from the medium or settles by gravity in the device's housing, and the gas passes from small pores of filter so it becomes clean.

The filtration process and bag-cleaning methods depend on the type of fabric filters, specifically, the composing materials and their resistance to acid or alkali gas attack and the structure geometry. Typically, the exit gas particulate concentrations of fabric filters are less than 5 mg/m³. The efficiency can reach 99% which depends on the type of filter and the condition of its operation. However, the efficiency of removal of acid gases is between 85 and 90% [14]. High collection efficiency can be obtained in the case of small particle size in the range of 0.1–0.01 μm [15].

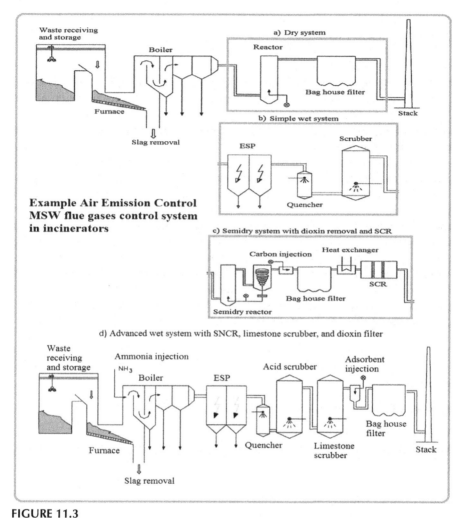

**FIGURE 11.3**

Illustration of the methods for flue gas system treatment rising from incinerators of MSW (Kristalina and Keshav, 1999).

An application of the abovementioned devices is given in Figure 11.3 which illustrates the working principles of the systems used for the control of air pollution produced by incinerators of MSW. The dry system and simple wet system (Figure 11.3a and b) can remove the particles of HCl, HF and heavy metals with medium control level of efficiency.

The semidry system with dioxin removal and the selective catalytic reduction (SCR), and the advanced wet system with the selective non-catalytic reduction (SNCR), limestone scrubber and dioxin filter (Figure 11.3c,d) can be used with advanced control level of efficiency to remove HCl, HF, $SO_2$, $NO_x$, dioxins and metals.

The selective non-catalytic reduction (SNCR) process at high temperature or a selective catalytic (usually $TiO_2$) reduction (SCR) at low temperature can be applied to remove $NO_x$. These are known as De-NOx processes. The De-NOx processes can convert nitrogen oxides to non-polluting nitrogen gas as $N_2$ and water vapor. The SNCR has better removal efficiency for the incineration plant, technically and economically, when ammonia and steam are injected into the furnace (Figure 11.3d).

It can be used as a catalyst to reduce gaseous pollutants by thermal incinerator systems (Nesaratnam and Taherzadeh, 2014).

In conclusion, a gas clean-up system depends on the type and quantity of waste in the gases effluent and on the emission limits to atmosphere, and on the treatment efficiency required. The finer fly ash particles can be collected from the flue gases by electrostatic precipitators (ESPs). Scrubber and bag house, respectively, can remove air pollutants and particulates from gas stream. In addition, electrostatic precipitation, fabric filtration, scrubbers and mechanical collectors or baghouses can remove most particulate control. In the modern incinerator systems, the efficiency of emissions control system complies with the environmental standards.

---

## Review Questions

What is the problem of air pollution?
What are the differences between recovery and destruction techniques of VOCs?

---

## References

1. General Environmental Regulations, Document No.1409-01, available online at www.pme.gov.sa, on 25/06/2020.
2. Royal Commission Environmental Regulations-2015, RCER-2015, Volume II, Environmental Permit Program, Environmental Protection and Control Department, available online at www.rcjy.gov.sa/en-us/riyadh/citizen/environment, on 25/06/2020.
3. P.T. Williams, 2005, "Waste Treatment and Disposal", 2nd Edition, Wiley, ISBN: 0-470-84912-6.
4. B. Bilitewski, G. Härdtle and K. Marek, 1994, "Waste Management", Translated and Edited by A. Weissbach and H. Boeddicker, Springer-Verlag Berlin Heidelberg GmbH, ISBN: 978-3-662-03382-1 (eBook).

5. F. Woodard, 2001, "Industrial Waste Treatment Handbook", Butterworth–Heinemann, ISBN: 0-7506-7317-6.
6. K.W. Harrison, R.D. Dumas, S.R. Nishtala and M.A. Barlaz, 2000, "A Life Cycle Inventory Model of Municipal Solid Waste Combustion", Journal of the Air and Waste Management Association Vol. 50, pp. 993–1003.
7. Y. Jin, L. Guo, O.D. Frutos, M.C. Veiga and C. Kennes, 2013, "Bioprocesses for the Removal of Nitrogen Oxides" in Christian Kennes and M.C. Veiga (eds.) "Air Pollution Prevention and Control: Bioreactors and Bioenergy", Wiley, ISBN: 9781119943310.
8. W.A. Worrell and P.A. Vesilind, 2012, "Solid Waste Engineering", 2nd Edition, Cengage Learning, Stamford, USA, ISBN-13: 978-1-4390-6215-9.
9. M.B. Hocking, 2005, "Handbook of Chemical Technology and Pollution Control", 3rd Edition, ISBN-13: 978-0-12-088796-5.
10. S.T. Nesaratnam and S. Taherzadeh (eds.), 2014, "Air Quality Management", Wiley, ISBN: 978 1 1188 6389 3.
11. K.B. Schnelle, Jr. and C.A. Brown, 2002, "Air Pollution Control Technology Handbook", CRC Press LLC. ISBN: 0-8493-9588-7.
12. K. Georgieva and K. Varma, 1999, World Bank Technical Guidance Report, "Municipal Solid Waste Incineration", The International Bank for Reconstruction and Development/the World Bank, Washington, DC, USA.
13. N. de Nevers, 2000, "Air Pollution Control Engineering", 2nd Edition, McGraw-Hill, ISBN: 0-07-116207-0.
14. D. Claffey, M. Claffey and J. Childress, 2014, "Fabric Filter Collectors" in Kenneth C. Schifftner (ed.) "Air Pollution Control Equipment Selection Guide", 2nd Edition, CRC Press, Taylor & Francis Group, ISBN-13: 978-1-4665-6182-3 (eBook).
15. L. Theodore, 2008, "Air Pollution Control Equipment", Wiley, ISBN: 978-0-470-20967-7.

# 12

## Economics of Waste Treatment and Management

### Key Learning Objectives

- Understanding the importance of cost on waste treatment and management.
- Understanding the cost of waste final disposal.
- Understanding how to estimate the cost of a waste project.

## 12.1 Introduction

The economic part is an important factor in any type of work that is related to public health and does not have direct recovery funds. However, collection, treatment and disposal of waste materials in an appropriate manner are considered necessary for the sake of public health.

The economic aspect of environment protection makes an essential part of any governmental development and economic budget. From an economic point of view, waste strategy in most of the world's countries focuses on achieving the zero-waste status in order to reduce waste materials and related services and costs.

There are several studies that deal with waste economy that includes evaluation, cost calculation and estimation of waste projects. Numerous engineering reports on waste management projects, such as wastewater and solid waste, are available in the archives of United States Environmental Protection Agency (EPA) [1]. Martinez-Sanchez et al. (2015) [2] presented an overview of the calculation principles and case studies for the economic assessment of solid waste management systems. Gamberini et al. (2013) [3] identified and analyzed engineering indices and costs for municipal solid waste management (MSWM). Aissa (2004) [4] presented an analytical analysis for the design of the wastewater treatment unit by comparing the mechanical costs at three levels of efficiency. A study by Visvanathan and Tränkler (2003) [5] showed that the total budget for municipal solid waste management in Sri Lanka is between 3 and 15% of the total budget of the local authority, 80% of which goes to collection and transport.

Waste management costs depend on several factors related to the type of waste, the amount of waste produced daily, the local economy of the society, the conditions of the site of treatment and other factors such as labor costs. Waste costs also include the costs of facilities (bins, machines and transport vehicles in case of MSW or infrastructure in case of wastewater), costs of waste operation and management. The cost is influenced by the adapted type of technology and methods of treatment.

## 12.2 Cost of Waste Minimization

Generation of waste materials is related to the level of civilization in the society and the economic level; however, this relationship is not linear. The annual costs for managing waste increases linearly with the increase of waste produced (Gamberini et al., 2013).

Waste minimization needs a governmental strategy and planning as explained in the waste prevention hierarchy (see Chapter 5). It needs time and financial support as well. Specifically, funds are needed for waste management processes such as recycling, reusing, energy recovery, landfill, incineration and composting of waste.

## 12.3 Waste Cost Estimation

Cost estimation is an important factor for any industrial project. Any profit or nonprofit project is subject to financing charges that depend on the capacity and size of the project. There are small, intermediate and large projects, all of which are linked to the economic assessment method. The cost estimate for waste management, as an industrial project, includes the capital of investment in treatment equipment and facilities, and the costs of operation and labor.

The total capital of investment can be divided into two parts: investment costs and operating expenses (Burstein et al., 1999 [6]). Investment costs include all costs necessary for the construction of a waste treatment plant. These include the costs of preliminary studies, project costs (design cost, engineering fees and legal applications permit), site preparation costs, acquisition and instrumentation costs, costs of construction and installation and start-up costs.

Investment costs are also known as Fixed-Capital Investment ($C_F$ or FCI). The FCI is the capital that is needed to supply the required manufacturing and plant facilities. Typically, it is divided into two types of costs:

a) Direct costs ($C_D$) that are related to manufacturing fixed-capital investment. The $C_D$ is the capital necessary for the installed process equipment with all the auxiliaries (components) that are needed to complete process operation. The $C_D$ includes the costs for the purchased equipment, installation, piping, instrumentation and control, electrical, site improvement and foundations, auxiliary insulation and land.

b) Indirect costs ($C_I$) are the costs related to nonmanufacturing fixed-capital investment. $C_I$ includes costs for plant components that are not directly related to the operation of the project (products or services) but are indirectly related to the performance of the project. The costs of processing buildings, off-site facilities (administrative and other offices, warehouses, laboratories and shops), engineering works, start-ups, contractor fees and contingency are included in $C_I$.

Thus, $C_F$ can be expressed as:

Fixed-capital investment = direct costs + indirect costs

Or;

$$C_F = C_D + C_I \tag{12.1}$$

The second part (work capital cost, $C_W$) of Equation 12.1 includes all operating and maintenance (O&M) costs, such as operating costs, maintenance costs, personnel, materials needed, energy, administration and overhead expenses, such as taxes and insurance.

The capital cost or the total cost ($C_T$) includes Fixed-Capital Investment ($C_F$) and working cost ($C_W$). This can be expressed mathematically as:

$$C_T = C_F + C_W = C_D + C_I + C_W \tag{12.2}$$

The cost depends on the specific characteristics of the technologies used, scale of the project and the costs of land (site).

## 12.4 Economic Analysis

Project cost analyses or profitability measures, such as Internal Rate of Return (IRR), Net Present Value (NPV) and Payback Period (PBP), are the standard profitability measures used in any industrial project. They are key to decision-making, design and policy strategy of management.

The common economic factors related to the duration of loans are the interest rate ($i$) and the number of years (n). The following are the main formulas used to calculate the interest factor.

$$\text{Simple payment amount (or future value FV); } F = P(1 + in) \tag{12.3}$$

$$\text{Single payment compound amount; } F = P(1 + i)^n \tag{12.4}$$

$$\text{Single payment present worth (or present value PV); } P = F(1 + i)^{-n} \tag{12.5}$$

$$\text{Compound amount (uniform series); } F = A[[(1 + i)^n - 1]/i] \tag{12.6}$$

$$\text{Sinking Fund payment; } A = F[i / [(1 + i)^n - 1]] \tag{12.7}$$

$$\text{Present Worth (uniform series); } P = A[[(1 + i)^n - 1]/[i(1 + i)^n]] \tag{12.8}$$

$$\text{Capital Recovery; } A = P[[i(1 + i)^n]/[(1 + i)^n - 1]] \tag{12.9}$$

The capital recovery factor can be used for annual cost analysis purposes.

---

## 12.5 Project Cash Flow

Cash flow analysis for a new project is illustrated in Figure 12.1. The figure shows the effect of rate of interest ($i$) during the project life time (n) (Smith, 2005 [7]); when $i = 0$ (curve I) and $i > 0$ (curves II and III), the value of "$i$" in the curve II is less than in curve III. It could be observed that in the case of curve II the profit is less than the maximum profit during the project life without interest, while in the case of high value of "$i$" (curve III), the profit of the project is zero at the end of the project life. The profit is measured by Net Present Value (NPV) while the Present Value (PV) is known as present discounted value [8].

Cash flow (CF) is the variation of money into and out of a business project in a time period, while the discounted cash flow (DCF) is an analysis method for valuing a project.

If the $A_j$ is the annual cash flow discounted at the end of year $j$ with the rate of interest "$i$", then the annual discounted cash flow $A_{DCF}$ is obtained at the end of year j. Thus:

at the end of year 1 => $A_{DCF1} = A_1/(1 + i)$;
at the end of year 2 => $A_{DCF2} = A_2/(1 + i)^2$
and at the end of year j => $A_{DCFj} = A_j/(1 + i)^j$

NPV of a project is the sum of the annual discounted cash flows ($\Sigma A_{DCF}$) over $n$, which is the years of project life (NPV = $\Sigma A_{DCF}$), that is directly dependent on $i$ and $n$.

Payback time or payback period (PBP) is the payback time from zero to date of cash breakeven point (Figure 12.1), in which the capital itself and the interest on it must be entirely paid back.

PBP is the fixed capital investment (FCI) over the annual cash flow ($A_j$) (PBP = $FCI/A_j$) or PBP is the total investment over the average annual cash flow. When PBP is a shorter time, then the project is more attractive for investment.

Return on investment (ROI) is expressed as a percentage; it is the net annual profit over the total investment multiplied per 100%.

An investment in a project starts at zero point (Figure 12.1) on the cumulative cash flow axis; if it goes to negative direction, money goes out from owner's pocket in stages. The first stage consists of project design, development and preliminary work. The second stage is the investment in buildings and equipment. The third stage of investment is the working capital needed when the plant is complete and ready to start; this indicates that the investment reached the maximum point.

At the initial production life of the project, it starts with a production capacity that is less than the designed full production capacity. The revenue from sales begins at the starting date of productive life, and the cumulative cash flow reaches again to zero on the cumulative cash flow axis (this is the project breakeven point). Over this point, the project starts to gain profits and it goes on along the whole project life. After that, the project can be still productive but there will be an increase in issues such as maintenance. The accumulation of cash flow over the time decreases (Figure 12.1) when it arrives the real end of project life. The final cash inflow entry from the project is the salvage value, that it is the resale value of an asset at the end of its useful life.

Figure 12.2 shows the cost of production and sales revenues. If the production quantity is less than the breakeven point (QB), then the project loses and while exceeding this point indicates that the project is profitable. The profit of the project is the gross profit or the total earnings. This can be expressed as:

Gross profit = total income – total product cost, (P = R – TC).

Where: P = Gross profit; R = Total income (Cash flows); TC = Total costs.

In general, the range of gross profits in industrial manufacturing is about 10–20%; it depends on the type of industry. In the food industry, the average gross margin is 37% while it is 57% in the beverage sector [9].

In production, the breakeven point is reached when total revenues (R) equal total costs (TC), or the profit zero. Usually, the net profit (NP) is achieved after the subtraction of depreciation charge: NP = R – TC – d;

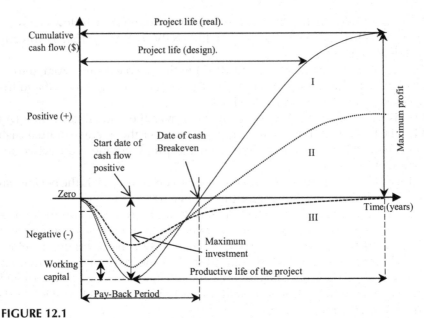

**FIGURE 12.1**

A typical cash flow for a new project.

where d = depreciation charge. Depreciation charge or expense is a percentage portion of the original cost that represents the periodical entries to complete the fixed expense of the property during its useful life. There are different methods to calculate depreciation; the simplest method is the linear method:

Depreciation per year = [Cost – 0 value] / Useful Life (years)

The net annual profit is obtained after the subtraction of all expenses such as depreciation and annual taxes:

Net annual profit = gross annual earnings – income taxes ($\tau$), (NP = R – TC – d – $\tau$)

For detailed information about the cost analyses, the reader can refer to specialized textbooks such as *The Engineer's Cost Handbook: Tools for Managing Project Costs* [10], *Handbook on Economic Analysis of Investment Operations* [11] and *Economic Analysis Handbook: Theory and Application* [12].

Several case studies were reported in the literature on waste management cost estimation. The US Department of Energy (DOE) [13] provided a case study of hazardous waste disposal costs based on a certain type of waste, waste location and transportation distance. For more details about hazardous waste treatment and cost estimation, refer to "Advances in hazardous industrial waste treatment" [14].

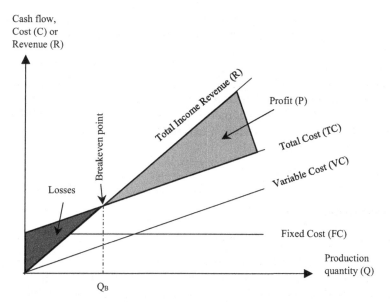

**FIGURE 12.2**

Minimum quantities of profitable production.

## 12.6 MSW Cost Estimation

Cost of MSW collection is estimated on the basis of the waste quantity (cost per ton). The cost of MSW collection is integrated in the waste management system. It includes all costs of collection services (equipment capital, operation and maintenance costs) such as collection vehicles, storage containers, personnel and the costs of transfer to treatment stations.

The life of the MSW project is usually more than 20 years (Dominic Hogg, 2011) [15]. The cost of the waste project depends on the type of project, such as landfilling, incineration, composting and mechanical biological treatment. The general formula for cost estimation of MSW is:

Cost ($) = waste generated quantity (tons) x percent of waste collected (%) x [cost of collection ($/ton) + cost of disposal ($/ton)]

Some indicative costs of waste treatment and disposal [16] are:

Landfills: 5–10 euro cents/kg;

Incinerators: 10–20 euro cents/kg;

Biological and chemical–physical treatment: 10–20 euro cents/kg;

Special wastes (wastes including construction and demolition (C&D), nonhazardous and hazardous wastes): 0.5–2.5 euros/kg.

It is obvious that the cost of landfilling disposal per ton is less than the cost of incineration disposal. For detail information about MSW cost estimation, the reader can refer to the EPA document: "The Full Cost Accounting for Municipal Solid Waste Management: A Handbook" (EPA, 1997) [17] and the Report of European Commission "Costs for Municipal Waste Management in the EU" (Dominic Hogg, 2011).

## 12.7 Wastewater Cost Estimation

In the case of wastewater, the cost includes the cost of piping infrastructure, the treatment plant and the cost of operation. Several examples of cost estimates for the waste projects as guidelines can be found in the US EPA website, e.g. "The Construction Costs for Municipal Wastewater Treatment Plants" which reports data collected from 1973 to 1978 (EPA document number 40980003) [18]. These data can be used to estimate new similar plants and to use time factors or cost indexes and scaling factor.

The typical scaling factor can be used to estimate the cost of the equipment or the entire plant. If the cost of an equipment or plant with its capacity is known, then the cost of a similar new one with its capacity could be estimated using the following equation:

$$C_A/C_B = (S_A/S_B)^n \qquad (12.10)$$

Where:

$C_A$ = cost for plant A

$C_B$ = cost for plant B

$S_A$ = size of plant A.

$S_B$ = size of plant B

$n$ = *cost-capacity factor* to reflect economies of scale.

Estimating cost by scaling is sometimes referred to as the exponential model. There are several values of $n$ on the basis of the type of equipment. In general, $n$ values range is between 0 and 1 for estimation of the equipment, or 0.5 and 1 for estimation of the plant. However, the $n$ value is usually used when n = 0.6. It is known as the "six-tenths-factor rule".

If the equipment increases in size, its cost also increases, but there is no linear relationship between capacity and cost. Taking in consideration the time factors, Equation 12.10 can be re-written as:

$$C_A = C_B(I_A/I_B)(S_A/S_B)^n \qquad (12.11)$$

Where: $I_A$ = index at the A time; and $I_B$ = index at the B time.

## 12.8 Pollution Cost Estimation

The cost estimate of the gas-cleaning project depends on the application of the process and the types of waste involved. In the case of MSW incineration, the cost of gas cleaning is included in the cost of incineration project. In the case of industrial gases effluent, the cost is included in the industrial project.

Cleanup costs are associated with chemical–physical parameters of gas stream and the technology applied. For detailed information about air pollution techniques and cost estimation, it is recommended to refer to "Fundamentals of air pollution engineering" [19], "Air pollution control technology handbook" [20] and "Air quality management" [21].

## 12.9 Income from Waste

Waste recovery as material or energy can reduce the overall cost of the waste treatment and disposal process. For example, the cost of MSW recycling includes separation equipment and tools, sorting and processing systems, maintenance and operation, enclosed buildings and containers for storage of prepared materials and worker support facilities. Biogas produced from landfill sites, if used, can reduce the transportation costs. Recycled materials markets can reduce the costs of separation and sorting. An estimation model for recycling revenue [22] is mathematically expressed as:

$$R = Q[R_w + rR_s - S_c - (1-r)D_c] \qquad (12.12)$$

Where:

R = revenue, gross income from waste ($);

Q = waste quantity (tons);

$R_w$ = income from receiving waste ($/ton);

r = recovery rate of waste, from original waste received (%);

$R_s$ = income from waste sales ($/ton);

$S_c$ = sorting cost and storage ($/ton);

$D_c$ = disposal cost for residual waste ($/ton).

Cost reduction can also be achieved by adapting pre-separation, such as the separation of plastic, glass and metal from the biowaste as composting feedstock; the composting costs will be reduced by about 10–20% (Luis Diaz et al., 2002 [23]).

## 12.10  Conclusion

Waste control and management is usually performed under the supervision of the authority of the country. In addition, waste management projects are considered to be major projects that need substantial amount of governmental financing or any third party. For this reason, some countries impose taxes on the goods or on the producing companies to cover environmental and disposal of waste costs. The government can, by imposing environmental and waste taxes, finance the cost of waste management.

## Review Question

Estimate the cost of waste treatment projects for a city, one for MSW and the second per wastewater treatment. Assuming the population of the town is 20,000 people, the MSW produced is 2.5 kg per capita and 200 liters of wastewater per capita. Both projects have 30 years of life and are located 50 km from city center.

## References

1. United States Environmental Protection Agency, www.epa.gov/; http://archive.epa.gov/.
2. V. Martinez-Sanchez, M.A. Kromann and T.F. Astrup, 2015, "Life Cycle Costing of Waste Management Systems: Overview, Calculation Principles and Case Studies", Waste Management, Vol. 36, pp. 343–355.

3. R. Gamberini, D. Del Buono, F. Lolli and B. Rimini, 2013, "Municipal Solid Waste Management: Identification and Analysis of Engineering Indexes Representing Demand and Costs Generated in Virtuous Italian Communities", Waste Management, Vol. 33, pp. 2532–2540.
4. W.A. Aissa, 2004, "Preliminary Design and Cost Estimation of Wastewater Treatment Unit", Eighth International Water Technology Conference, IWTC8 2004, Alexandria, Egypt.
5. C. Visvanathan and J. Tränkler, 2003, "Municipal Solid Waste Management in Asia: A Comparative Analysis", Workshop on Sustainable Landfill Management, 3–5 December; Chennai, India, pp. 3–15.
6. D. Burstein et al., 1999, "Environmental Engineering" in R.A. Corbitt (ed.) "Standard Handbook of Environmental engineering", 2nd Edition, McGraw-Hill, ISBN: 9780070131606.
7. R. Smith, 2005, "Chemical Process Design and Integration", Wiley, ISBN: 0-471-48680-9.
8. Wikipedia, Present value, available online at https://en.wikipedia.org/wiki/Present_value, on 24/9/2015.
9. Azcentral.com, Gross Profit Margin in Food Industry, available online at http://yourbusiness.azcentral.com/gross-profit-margin-food-industry-26284.html, on 25/9/2015.
10. R.E. Westney (ed.), 1997, "The Engineer's Cost Handbook: Tools for Managing Project Costs", Westney Consultants International, Inc. Houston, Texas, Marcel Decker, Inc. ISBN 0-8247-9796-5.
11. P. Belli, J. Anderson, H. Barnum, J. Dixon and J.-P. Tan, 1998, "Handbook on Economic Analysis of Investment Operations", Operations Policy Department Learning and Leadership Center of The World Bank. (Pdf, Download from managingforimpact.org. Available online at www.managingforimpact.org/resource/world-bank-handbook-economic-analysis-investment-operations or UNDP, www.undp-alm.org/resources/relevant-reports-and-publications/handbook-economic-analysis-investment-operations).
12. A.D. Stament, W.H. Bennett and W.S. Moore, 1973, "Economic Analysis Handbook: Theory and Application", Volume 2, Concepts and Techniques, General Research Corporation, Operation Analysis Division. Westgate Research Park, McLean, Virginia 22101. Distributed by National Technical Information Service (NTIS), U.S. Department of Commerce.
13. DOE, Department of Energy, document number G 430.1–1 (03-28-97), "Cost Estimating Guide", U.S. Department of Energy, Associate Deputy Secretary for Field Management.
14. L.K. Wang, N.K. Shammas and Y.-T. Hung (eds.), 2007, "Advances in Hazardous Industrial Waste Treatment", CRC Press Taylor & Francis Group. ISBN-13: 978-1-4200-7230-3.
15. D. Hogg and Eunomia Research & Consulting, 2001, "Costs for Municipal Waste Management in the EU", Final Report to Directorate General Environment, European Commission.
16. E. Carradori, L. Cutaia and G. Mastino, 2009, "Industrial Waste Management: Environmental and Economic Impact of Waste Produced by Major Accident Hazard Industrials", Prevention Today Vol. 5, Supplement 3/4, pp. 53–72.

17. EPA, document number 530-R-95-041, 1997, "Full Cost Accounting for Municipal Solid Waste Management: A Handbook", EPA, USA.
18. EPA, document number 430/980003, April 1980, Technical report (FRD 11): Construction Costs for Municipal Wastewater Treatment Plants: 1973 to 1978. Available online at www.kysq.org/docs/Wastewater_1980.pdf.
19. R.C. Flagan and J.H. Seinfeld, 1988, "Fundamentals of Air Pollution Engineering", Prentice-Hall, ISBN 0-13-332537-7.
20. K.B. Schnelle, Jr. and C.A. Brown, 2002, "Air Pollution Control Technology Handbook", CRC Press, ISBN 0-8493-9588-7.
21. S.T. Nesaratnam and S. Taherzadeh (eds.), 2014, "Air Quality Management", John Wiley & Sons Ltd, in association with: The Open University, United Kingdom. ISBN: 978-1-1188-6389-3.
22. rebri.org, reducing building material waste, 2014, "Resource Recovery—All Waste Types—Centralised Sorting and Storage", available online at www.rebri.org.nz
23. L.F. Diaz, G.M. Savage and C.G. Golueke, 2002, "Composting of Municipal Solid Wastes" in George Tchobanoglous and Frank Kreith (eds.) "Handbook of Solid Waste Management", 2nd Edition, McGraw-Hill, New York.

# *Index*

Printed in the United States
By Bookmasters